# HIGH-TECHNOLOGY WORKPLACES

## Integrating Technology, Management, and Design for Productive Work Environments

*Edited by Pierre Goumain*

 VAN NOSTRAND REINHOLD
_____ New York

T
59.77
.H54
1989

The following chapters have been reprinted with permission:

Chapter 5, "The Role of Land and Facilities in Fostering Linkages Between Universities
and High-Technology Industries," originally appeared in a publication by the
Society for College and University Planning in 1985.

Chapter 13, "Cable Management in the High-Technology Workplace," originally
appeared in *Facilities,* vol. 3, no. 6, June 1985.

Printed in the United States of America

Van Nostrand Reinhold
115 Fifth Avenue
New York, New York 10003

Van Nostrand Reinhold International Company Limited
11 New Fetter Lane
London EC4P 4EE, England

Van Nostrand Reinhold
480 La Trobe Street
Melbourne, Victoria 3000, Australia

Macmillan of Canada
Division of Canada Publishing Corporation
164 Commander Boulevard
Agincourt, Ontario M1S 3C7, Canada

16  15  14  13  12  11  10  9  8  7  6  5  4  3  2  1

Library of Congress Cataloging-in-Publication Data

High-technology workplaces.

Bibliography: p.
Includes index.
1. Work environment.  2. Facility management.
3. High technology industries.  I. Goumain, Pierre.
T59.77.H54  1989        658.4'04        88-27674
ISBN 0-442-22741-8

# Contents

# Preface

**G**iven the subject matter of this book, it is somewhat ironic that while working intensively on my word processor at home to complete it, I managed to get a stiff neck that was very sore for several days! I blamed both my need for bifocals and my chair with insufficient back support. I could thereby appreciate, if that's the right word, the need for an ergonomic workstation at home and experience firsthand some of the problems of high-technology workplaces.

The workplace is changing rapidly as a result of the introduction on an unprecedented scale of new information technologies in organizations. These technologies have an impact on all economic and sociocultural aspects of industrially developed societies. As a consequence, work processes and the social environment of organizations are being redefined. At the same time, the physical environment of the workplace must respond to the new environmental and organizational needs arising. There is now an emerging body of knowledge and experience that enables us to tackle the design and management problems of high-technology workplaces in an integrated way. This book takes stock of what is known today and presents approaches to the problem from different perspectives. It is not a monograph of high-technology workplaces, and examples are given in the text mainly to illustrate salient points made in the text or support the discussion of relevant issues.

The progress of work automation has been very fast during the past decade, its pace sometimes outstripping the capacity of organizations to absorb new technologies and to reap their potential benefits effectively. Initially, it primarily concerned heavy manufacturing industries, affecting manual work and the jobs of blue-collar workers. It has now spread into light manufacturing and office work, with an impact on white-collar workers, technicians, and the knowledge workers. This book focuses on the latter.

With the merger of computing and telecommunication industries and media, which is increasing computing power while decreasing costs, most occupations will soon have been affected by the ubiquitous new technologies. And so will most workplaces. So that in many respects this book also addresses the present or future needs of most workplaces.

As the use of high-technology products and services increases, high-technology industries are acquiring greater economic significance. In an increasingly competitive business climate, it is vital for most organizations in the private and the public sectors to be successful in managing the introduction of technological innovations and corresponding organizational transitions, or transmutations. These changes sometimes imply a reorganization of work processes that transgresses traditional work categories, such as "production" or "office," and corresponding building types. The fostering of productive work environments for high technology, therefore, must deal with both the production activities for products and services and the use of these in organizations. Work settings have a role to play in both respects, and this book tackles both.

With regard to the second aspect, high technology is affecting the workplace in two different and complementary ways. A direct effect is that the workplace is being invaded by a whole new range of equipment with new, and frequently conflicting, environmental needs. An indirect effect is that workplaces must change to adapt to the transformations of work processes and organizations that result from automation. This book considers both effects.

Because the impacts of information technology have many systemic ramifications, some of which have implications for the physical work environment, this book favors a holistic approach that looks at the technological, social, and environmental aspects together. Its emphasis is on the *integration* of technology, management, and design for productive work environments, a theme that runs throughout all its chapters. In this view, the physical environment is seen as a tool of management that contributes to physical and psychological wellness, productivity, and the accomplishment of organizational objectives. The challenge for management is to maintain a fit between rapidly changing technologies and work processes, and people, organizations, and physical settings. The physical workplace, then, is considered as an integral part of managerial policy and business plans, an investment that yields returns, rather than just a necessary cost to be depreciated over a number of years. Thus, high-technology workplaces are becoming high-quality workplaces, both in terms of their environmental appropriateness and their responsiveness to change.

The initial impetus for this book came from a conference I organized in March 1985, while a faculty member of the Department of Architecture,

State University of New York at Buffalo. In doing so I was fortunate to receive start-up support from the Conferences in the Disciplines Funding of the University.

Other U.S. organizations also lent their support, including: The American Institute of Architects, Western New York Chapter; The Amherst Industrial Development Agency; The Buffalo Area Chamber of Commerce; *Business Facilities / High Tech Facilities* magazine; The Erie County Industrial Development Agency; The National Association of Industrial and Office Parks; The New York State Association of Architects Inc./The American Institute of Architects; The New York State Economic Development Council Inc.; The New York State Science and Technology Foundation; and The Technical Societies of the Niagara Frontier.

A number of speakers at the conference also contributed chapters to this book, including: Colin Davidson, Gerald Davis, David Eakin, Harry Eggink, Ira Fink, Scott Lambert Gardiner, David Levy, Cecil Williams, Charles Minshall, and John Worthington. Other speakers included: Jim Allen, Michael Brill, Michael Brooks, Graham Jones, Norman King, Arthur Moog, Bert Orr, James Ridler, Steven Sample, Robert Shibley, and Douglas Stoker. Particular thanks are due to Richard Rogers, who produced a stimulating presentation at the conference of recent work on high-technology projects, and agreed to make a second and different presentation to all the students and faculty of the SUNY Department of Architecture.

Other people who helped toward the conference include: Dorothy Collins, Wayne Hazard, Ruth Klein, Lis Liu, Terry Moynihan, Sid Peters, Eric Peterson, Barbara Rodriguez, John Shaflucas, Richard Swist, Paul Van Wert, and Andrew Weil, as well as my students. I am particularly indebted to Jim Allen, without whose constant and enthusiastic support this project would never have been seen through, and to Elizabeth Minklei who, as research assistant, contributed invaluably. Bruce Campbell produced creative graphic designs for the conference flyer and the book cover (Is it a chip? Or is it a building?).

I continued work on this project as research associate of the GAMMA Institute, and visiting scientist at the Canadian Workplace Automation Research Centre (CWARC), Communications Canada, and more recently also as president of DesignErgo Inc. I particularly appreciated the support of Michèle Guay, director of organizational research at the CWARC, who shares much of the underlying philosophy integrating organizational and environmental aspects of high-technology workplaces. I am also indebted to René Guindon, director general, CWARC, and to Jacques Lyrette, until recently executive director, research, Communications Canada, as well as to Kimon Valaskakis, president of GAMMA, for their support.

Chapter 12, "Cable Management in the High-Technology Workplace," by Paul Stansall and Michael Bedford, is a revised version of an article by the same authors entitled "Cable Management Approaches," which appeared in *Facilities,* vol. 3, no. 6, June 1985, pp. 10–15.

Finally, grateful thanks are also due to Judy Joseph, Lilly Kaufman, Paul Lukas, Dorothy Spencer, Cindy Zigmund, and the staff of Van Nostrand Reinhold, without whose judgment, assistance, and patience this book would not have been completed.

# Contributors

**Pierre Goumain** is President of DesignErgo Inc., a Montréal-based research and practice company focusing on workplace environmental design and management. An architect with extensive experience in practice, research, consulting, and teaching in Europe and North America, Mr. Goumain is currently a Visiting Scientist with the Directorate of Organizational Research at the Canadian Workplace Automation Research Centre in Laval, Québec. Previously, he was a key member of the Department of Design Research at the Royal College of Art in London for over a decade, where he worked in such areas as design theory, design methods, programming and management, human/computer interaction in architectural design, and multi-purpose industrial developments. Mr. Goumain has written numerous research reports and publications, and has made contributions to many international conferences. He is a member of the Environmental Design Research Association in North America, and an International Council member of the Design Research Society in the United Kingdom.

**Franklin D. Becker** is professor of Human/Environment Relations and Facility Planning and Management at Cornell University, Ithaca, N.Y., and a principal in Facilities Research Associates, Inc., a consulting firm based in Ithaca, N.Y. His current research focuses on facility innovation and organizational performance. Dr. Becker received his doctorate in environmental and social psychology from the University of California at Davis in 1972. His publications include three books: *Housing Messages, Workspace: Creating Environments in Organizations,* and *The Successful Office.*

**Michael Bedford** is an Associate of DEGW, Architects, Planners & Designers, in London, where he is responsible for space planning and strategic studies for both public and private sector clients. Prior to joining DEGW, he practiced architecture in London and was research and energy coordinator at The Royal Institute of British Architects. At University College London, where he studied architecture, Mr. Bedford was involved in the Space-Syntax Research Project. Mr. Bedford is European editor of the *Journal of Architectural and Planning Research.*

**Colin H. Davidson** is full professor at the School of Architecture, University of Montréal, where he began teaching building industrialization in 1968. He has worked for the Architects' Collaborative in Cambridge, Mass., and for the London County Council—predominantly in the field of housing. Recently, Mr. Davidson's interests have included studies of the impact of the computer on design decision making in the building process and the role of incubators in local economic development. Mr. Davidson studied at the Brussels City Royal Academy School of Architecture and Massachusetts Institute of Technology, where he earned a master's of architecture.

**Gerald Davis** is president and chief executive officer of the International Center for Facilities, Seattle, Wash., U.S.A., and Ottawa, Ont., Canada. Having founded TEAG—The Environment Analysis Group—in 1965, he is a pioneer in the field of facility predesign, programming, and evaluation. He has worked at every scale, from urban design and planning to furnishings and computer equipment. Currently Mr. Davis is the chair of ASTM Subcommittee E06.25 on the Overall Performance of Buildings, and he was project director of ORBIT-2, a major multiclient research and consulting project about offices, information technology, and the organizations that use them.

**David B. Eakin** is the advanced-technology program engineer for the design and construction of federal buildings by the U.S. General Services Administration. In this capacity, he is responsible for the direction and development of design criteria and policies relating to innovative building technologies. A graduate of the University of Maryland, Mr. Eakin is a registered mechanical engineer. He is vice chairman of the Building Thermal Envelope Coordinating Council, and a member of ASHRAE and ASTM.

**Harry A. Eggink** is professor of architecture and urban design at Ball State University, Muncie, Ind. He holds a master's of architecture/urban design from Harvard University and was a Fulbright Scholar to Finland. In his professional career, he has done work for architectural firms in Germany, Washington, D.C., Boston, and Indianapolis. Mr. Eggink has won several architectural and urban design awards and has had several projects and studies published.

**Ira Fink** is president of Ira Fink and Associates, Inc., University Planning Consultants of Berkeley, Cal. For the 11 years prior to establishing his private practice in 1978, he was the university community planner for the office of the president of the nine-campus University of California system. He has also served as a planning consultant for many other universities across the country. Dr. Fink is a licensed architect; among his publications are an article on new university research and development parks and an article on the university as land developer.

**W. L. (Scot) Gardiner**, Ph.D., is an associate professor in the Department of Communication Studies at Concordia University and vice president of the GAMMA Institute in Montreal. His current specialization is the human impact of the new information technologies. Dr. Gardiner studied psychology at Cornell University and has published three textbooks in that field, *Psychology: A story of a Search, An Invitation to Cognitive Psychology,* and *The Psychology of Teaching.*

**Robert J. Koester** serves as director of the Center for Energy Research/Education/Service (CERES). He is a licensed architect and is professor of architecture in the College of Architecture and Planning at Ball State University, Muncie, Ind. With Harry Eggink, he won first prize in a national design contest sponsored by the National Trust for Historic Preservation. Mr. Koester received his graduate degree at Rensselaer Polytechnic Institute and did postgraduate studies at Harvard and MIT.

**David Levy** is a program manager with the Herman Miller Research Corporation in Ann Arbor, Mich., where he conducts product development research. He was previously on the architecture faculty at the University of Michigan, where he taught courses in design, building programming, and facilities management. Since 1978, Dr. Levy has been conducting research on the rehabilitation of industrial buildings; he has also done research and consulting in the area of health-care planning and programming.

**Charles W. Minshall** is program manager for community and regional development at Battelle, Columbus, Ohio. In his 15 years there, he has managed or served as principal investigator in more than 140 studies in the United States and 17 foreign countries. His research interests include urban and regional economic growth, environmental impact analysis, and the spacial interactions of individuals and firms. In addition, he has been involved in action-oriented research, such as that on inner city industrialization. Dr. Minshall earned a doctorate in geography from Ohio State University.

**T. J. Springer** is president of Springer Associates, an ergonomic consulting group in St. Charles, Ill. that assists businesses in addressing the issues encountered in the planning, design, and management of high-technology workplaces. The group's clients range from international Fortune 500 companies to small regional firms, and have included IBM, Pacific Bell, Monsanto, and Conrail. Dr. Springer has recently become director, Facilities Management Program, F.E. Seidman School of Business, Grand Valley State University. Dr. Springer, who holds a doctorate in human factors psychology, is widely published and has recently completed the book *Improving Productivity in the Workplace: Reports From the Field.*

**Paul Stansall** is an Associate of DEGW, Architects, Planners & Designers in London, where his main responsibilities are for facilities programming, strategic space planning, and building evaluation for corporate clients. He was recently involved in the ORBIT-2 research project, which examined the suitability of North American office buildings to meet increasing demands for information technology in the workplace. A member of the Royal Institute of British Architects, Mr. Stansall is currently a PhD candidate at University College London, where he is studying the spatial organization of corporate offices.

**Fred A. Stitt** is a California architect whose professional work is primarily researching and writing and teaching about innovative solutions to the problems of architectural practice. He writes and publishes such periodicals as "The Guidelines Letter," a monthly newsletter on innovations and advanced techniques in architectural and engineering practice, "Working Drawings—The Production Newsletter," "Solutions—The A/E Macintosh User's Newsletter," and "Marketing Guidelines." Among his recent books are *Designing Buildings That Work, Systems Graphics and Systems Drafting, The Design Office Management Handbook,* and *The Design Office CADD Handbook.*

**Cecil L. Williams** is the corporate director of Health and Wellness for Herman Miller, Inc., Zealand, Mich. Prior to this appointment, he worked for Facility Management Institute, a Herman Miller subsidiary. Dr. Williams has been director of the University Counseling Center at Michigan State University, and was also professor of counseling psychology there. In the past 10 years, his experience has led him to use Carl Jung's psychology in his work on individual personal styles in management and leadership roles. Dr. Williams's most recent publication (coauthored with Armstrong and Malcolm) is *The Negotiable Environment: People, White-Collar Work and the Office.*

**John Worthington** is a founding partner of DEGW, Architects, Planners & Designers, in London, a firm that specializes in matching organizational requirements to buildings. He has been responsible for space planning projects for a variety of firms, including Digital Equipment Corporation and Hewlett Packard, and has undertaken planning projects for the Scottish and Welsh Development Agencies. Currently, Mr. Worthington is involved with user research and programming for Stockley Park, Heathrow, London's first international business park.

CHAPTER ONE

# CHANGING ENVIRONMENTS FOR HIGH-TECHNOLOGY WORKPLACES

*PIERRE GOUMAIN*

The "office of the future" is here today, and here to stay. And not just the office; indeed, most workplaces. Most organizations have been affected by information technology. Soon, all will have been affected. They may be producing new technologies—products (hardware and software) and services—or they may be using these. Or they may be both producing and using them. Managers and office technology task forces are beginning to accept responsibility for new technologies and their impact on organizations and their social environment.[1] An awareness of the concomitant impact on the physical environment of the workplace is also becoming more widespread. The social and the physical environments are changing concurrently and systemically. For many organizations, managing such changes successfully is crucial. However, the scope and nature of changes due to the implementation of advanced information systems are unprecedented, resulting in complex organizational transitions, or even transmutations.

We are only beginning to come to grips with such changes. While we have largely coped with change in a reactive mode hitherto—putting out fires whenever and wherever intolerable environmental misfits appeared—a relevant body of knowledge and experience is developing that allows us to adopt a more proactive posture toward planning and managing change. This implies a closer integration of technology, management, and design. This book focuses on such issues and presents an emerging understanding of high-technology workplaces and their specific environmental problems.

In this introductory chapter, I will first outline the interaction between high technology and workplaces, then introduce the goals of this book and its intended audience. The structure of the book will then be presented, together with an overview of its contents and of major issues, followed by a discussion of changing environments in high-technology workplaces and some concluding remarks.

## HIGH TECHNOLOGY AND WORKPLACES

It has often been said that the current information revolution is comparable in scope, if not in kind, to the first industrial revolution: brain power today, as opposed to brawn power last century.[2] The impacts of information technology (IT) are systemic, affecting both production and consumption in all economic and sociocultural aspects of our lives in industrially highly developed countries.

### Producing High Technology

Internationally, high technology is increasingly perceived as essential to economic survival in an increasingly competitive world where new nations are challenging the hegemony of long-established economic powers. And at a national level, the regions of many countries are vying for high-technology development, both nurturing "homegrown" innovations and commercial applications and competing to attract industries from outside.[3] While high technology per se contributes only a relatively modest proportion of the gross national product and employment directly in industrialized countries, high-technology products and services have a ubiquitous effect that changes most work processes and organizations radically, and therefore also the workplaces that give them shelter. Conversely, workplaces may help or hinder production. For most of our buildings in use today were designed at a time when it was impossible to foresee—let alone design for—recent technological advances. Furthermore, traditional classifications of types of workplaces and activities are becoming less relevant or clearly defined[4]—particularly among the high-technology industries themselves:[5] For instance, innovation (research lab), production (factory), and control (office), once clearly subdivided, are merged again into total innovation and production systems through integrated information systems such as computer-aided design/computer-aided manufacturing (CAD/CAM).[6] And the "services" sector is not nearly as clearly defined as it used to be. As observed by the CALUS study on property and information technology: "Changes in the structure of the economy have resulted in a breaking down of the traditional distinction between office and non-office activities and a growing demand for more flexible buildings which are interchangeable between office, assembly, research and other activities."[7]

## Using High Technology

Economic health in the high-technology sector implies a growing market for its products and services. Apart from robotics and the application of IT to heavy manufacturing processes—an area which is outside the scope of this book—a major growth sector is office automation (OA) and light manufacturing. It has been only during the eighties that OA has come into its own as a result of such combined factors as an exponential growth in available computing power, a corresponding drop in prices, and the merger of computing and telecommunication technologies and media. Today, successful automation has become a matter of vital importance to most organizations. It also has become a challenge. The early accelerating rate of take-up of OA has slowed down more recently: following early promises and some dramatic successes, it became clear that the implementation of sophisticated OA systems affects productivity and work environments—both social and physical—in many complex and interrelated ways, and that some equally dramatic failures could be attributed to neglecting to take such factors sufficiently into account.

There has also been a growing realization that the physical environment of the workplace can no longer be perceived as a neutral backdrop to work processes: it is rather a catalyst of organizational change, a tool among others contributing to the accomplishment of organizational objectives. Thus, the physical environment is—or should be—part and parcel of the business plan of any organization.

## The Economics of High-Technology Workplaces

This realization implies a considerable shift in understanding and attitude, a shift that has begun to take place only recently. With increasingly challenging competition and scarcer resources, it has become necessary to make all resources—including facilities—contribute to an organization's mission as effectively as possible. However, to many organizations land, buildings, and equipment are still no more than necessary costs to be depreciated over a number of years rather than assets and investments contributing to organizational productivity and employees' wellness.

It is not always appreciated that cumulatively land, buildings, and equipment constitute some 69% of the wealth of the United States, and frequently over 25% of a company's assets.[8] With the advent of information technology, we are confronted with the task of coordinating different rates of obsolescence and depreciation for the IT hardware and software (fast), and the various elements of the physical environment ranging from furniture (medium) to buildings (slow).[9]

But perhaps most relevantly in connection with the subject matter of this book, we should emphasize the very substantial leverage potentially afforded by the physical environment in improving an organization's performance, hence its bottom line: Over a 10-year period, it is estimated that costs will be around 85% for personnel, 7% for facilities and maintenance, and 5% for equipment, with 3% for miscellaneous items.[10] Thus, an additional marginal investment in facilities can yield considerable returns in cost savings in personnel through improved health, productivity, and performance and increased job satisfaction and decreased absenteeism. According to Springer, with returns of 10 to 15%, which have been observed in some instances, payback periods can be as short as 1 yr.[11] Similar findings are reported in the Buffalo Organization for Social and Technological Innovation (BOSTI) study on office design and productivity.[12]

Also, with the high churn rates for organizational and physical (layout) changes that are increasingly common, and in part caused by IT directly and indirectly, the flexibility of facilities and equipment to respond efficiently and to minimize downtime (hence unproductive work time) also makes a contribution to productivity and to the bottom line. Finally, a more indirect and difficult to measure—but no less important—aspect of the economic performance of facilities is the role they may play in attracting and retaining the knowledge-workers essential to high-technology activities,[13] either in direct connection with their work mission[14] or more general life-style,[15] in fostering workers' psychological wellness and in conveying an appropriate corporate image to actual and potential customers and to the general public.

## GOALS AND INTENDED AUDIENCE

It has become clear that to truly reap the potential benefits of high technology we need to find ways of integrating technology, management, and design to facilitate organizational transitions and to foster productive work environments in high-technology workplaces. It is critical to maintain a fit between changing information and telecommunication technologies, changing work processes, changing organizations and people, and changing physical environments. This book is a contribution toward such goals.

Clearly the accomplishment of such goals entails the transgressing of traditional disciplinary and professional boundaries. Just as the design and implementation of IT systems often provoke the recasting of organizational structures and work systems to ensure that the master/slave relationship between people and technology is not reversed inadvertently,[16] so too do they indirectly provoke a reassessment of role, responsibility, and authority for the various players involved in the processes intended to deliver productive work environments.

Thus, this book is intended for all the decision makers who contribute to the integrated technology/management/design processes for high-technology workplaces at various levels, from strategic planning to day-to-day facility management. The theme of integration of these three elements runs throughout this book.

People who may benefit from this book include: politicians, corporate planners, and decision makers and managers concerned with local economic development in relation to high-technology activities; university leaders concerned with the role of universities in local economic development and with the role of facilities in the attainment of academic and research excellence; corporate and government managers and facility programmers and managers involved in new or reuse projects; and the various participants in the construction industry, both the traditional participants (architects, interior designers, engineers, developers, builders, equipment and furniture designers and manufacturers, etc.) and the relative newcomers involved in the applications of information technology to building systems. In other words, all those who contribute to the facility programming, design, and construction management and the post-occupancy evaluation (POE) of high-technology workplaces.

It may be argued that all members of such a potentially wide audience cannot be served with sufficient depth in just one book. However, the goals of this book, as stated above, call for a synthetic view, for the reader to *integrate* technology, management, and design, and to begin to weave a web of connections between relevant topics for this new subject area. Otherwise, and as suggested by Lambert Gardiner for his model in the next chapter, we may be "like the three blind men, each in contact with a part of the elephant and getting a false view of the whole elephant."

Before an overview of the structure and contents of this book is presented, two key terms should be defined. The first, *design,* in the context of this book should be understood in its widest sense to encompass the whole cycle of activities leading to physical changes, including the identification of human needs and the production of programmatic information, alternative design (in its narrower, more traditional sense), procurement, and construction methods, and the continual post-occupancy evaluation, diagnostic, and management process. The second term, *high-technology workplaces,* is not intended to refer to what architects have come to recognize as the "high-tech" architectural style: the sleek "building as machine" aesthetics of late-modernist functionalism as epitomized by the widely known Pompidou Center in Paris. While such a style may be appropriate to the corporate image projected by some high-technology companies, its use cannot in and of itself ensure the design of work environments appropriate to a wide variety of high-technology work processes.

## OVERVIEW

As shown in Figure 1-1, this book comprises 15 chapters, excluding this introductory chapter. They have been organized into four sections, moving from the general to the specific. Part I deals with societal trends, Part II with the planning aspects of developments for high technology, Part III with aspects of high-technology projects, and Part IV with specific examples.

### Part I: Societal Trends and High Technology

Since much material can be found elsewhere about the economic and sociocultural impacts of the information revolution, this section serves only as context and backdrop to the book as a whole. W. Lambert Gardiner's Chapter 2 introduces major trends in the information society—thus adding to the rapidly growing "metatrend" of trend analyses—and discusses issues of forecasting and planning in a lively way while helping us to form broad concepts to map out complex societal trends. In his model, which consists of the overlapping ecosphere, sociosphere, and technosphere, he suggests that information technology is a prime mover in the currently predominant technosphere-as-cause scenario. Our capacity to forecast and to plan—and thus to control to some degree—this trend depends on our subjective perceptions of this basic model. Distinctions between basic activities such as learning, working, and playing are becoming blurred. In the future we may lead more balanced lives, with time devoted to each activity distributed more evenly than it is in the successive education, work, and retirement phases of our present lives. We may see increasingly diverse "convivial environments" designed for such activities.

In Chapter 3, John Worthington starts from an examination of broad societal changes to discover ways of accommodating the knowledge-based industries: changes in patterns of employment (increased use of brain power), in technologies that are increasingly convergent, in methods of production (speed of innovation and change, shorter product life spans, global products), and in patterns of work (toward batch production, overlapping engineering design and manufacturing, flexibility of time and

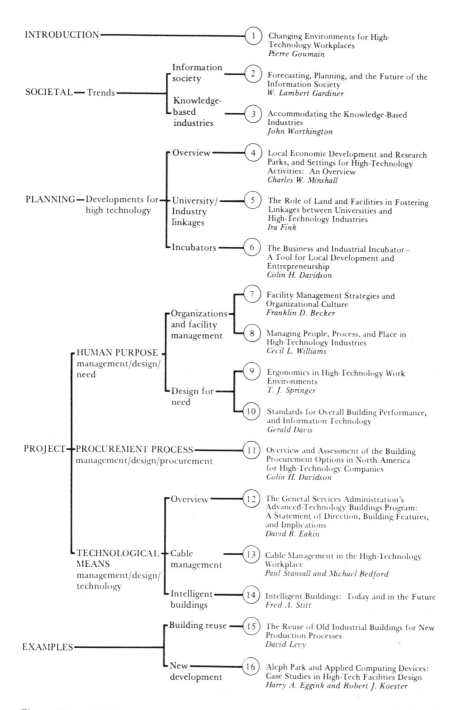

INTRODUCTION ———————————————— (1) Changing Environments for High-
Technology Workplaces
*Pierre Goumain*

SOCIETAL — Trends
- Information society — (2) Forecasting, Planning, and the Future of the Information Society
*W. Lambert Gardiner*
- Knowledge-based industries — (3) Accommodating the Knowledge-Based Industries
*John Worthington*

PLANNING — Developments for high technology
- Overview — (4) Local Economic Development and Research Parks, and Settings for High-Technology Activities: An Overview
*Charles W. Minshall*
- University/Industry linkages — (5) The Role of Land and Facilities in Fostering Linkages between Universities and High-Technology Industries
*Ira Fink*
- Incubators — (6) The Business and Industrial Incubator— A Tool for Local Development and Entrepreneurship
*Colin H. Davidson*

PROJECT
- HUMAN PURPOSE management/design/need
  - Organizations and facility management
    - (7) Facility Management Strategies and Organizational Culture
*Franklin D. Becker*
    - (8) Managing People, Process, and Place in High-Technology Industries
*Cecil L. Williams*
  - Design for need
    - (9) Ergonomics in High-Technology Work Environments
*T. J. Springer*
    - (10) Standards for Overall Building Performance, and Information Technology
*Gerald Davis*
- PROCUREMENT PROCESS management/design/procurement — (11) Overview and Assessment of the Building Procurement Options in North America for High-Technology Companies
*Colin H. Davidson*
- TECHNOLOGICAL MEANS management/design/technology
  - Overview — (12) The General Services Administration's Advanced-Technology Buildings Program: A Statement of Direction, Building Features, and Implications
*David B. Eakin*
  - Cable management — (13) Cable Management in the High-Technology Workplace
*Paul Stansall and Michael Bedford*
  - Intelligent buildings — (14) Intelligent Buildings: Today and in the Future
*Fred A. Stitt*

EXAMPLES
- Building reuse — (15) The Reuse of Old Industrial Buildings for New Production Processes
*David Levy*
- New development — (16) Aleph Park and Applied Computing Devices: Case Studies in High-Tech Facilities Design
*Harry A. Eggink and Robert J. Koester*

**Figure 1-1.** Structure of the contents of the book.

place to work, job sharing, telecommuting). For high-technology companies, the accelerating pace of growth and change moves through successive stages, from embryonic to institutional, with different environmental
needs at each stage. At the same time, increasingly diverse and constantly
changing functions and departments are housed under one roof. Building
shells must be designed with sufficient flexibility to respond to such changing and diverse needs.

## Part II: Planning Developments for High Technology

At a planning level, we can observe the need to integrate management and
design to facilitate developments that will attract, nurture, and retain
high-technology activities, an issue that has become very relevant to many
competing local economic development agencies, as mentioned earlier.
Thus high-quality work environments and settings now are a *sine qua non*
for high-quality knowledge workers. In Chapter 4, Charles W. Minshall
gives an overview of research parks and settings for high-technology activities, with a number of criteria that enable us to discriminate among the
many varieties and to identify appropriate conditions for success, among
which two stand out: the use of formal and clearly documented covenants
and local university support. The latter is the focus of Ira Fink's Chapter 5.
Both Minshall and Fink observe the increasing mix of diverse activities, including many service facilities, in most successful parks, while Minshall
emphasizes the need for managerial leadership.

In Chapter 5, Ira Fink focuses on the critical role of land and facilities in
fostering linkages between universities and high-technology industries.
This has gained momentum in the United States in recent years, with
government—federal, state, and local—as the third partner. As the chief
creators of new knowledge, universities have a major role to play for
knowledge-based industries and local economic development through
technology transfer. With this trend toward "academic capitalism," two
groups are emerging within universities; the "gatekeepers of traditional
values and the entrepreneurs." Fink presents several types of research
linkages (consulting, research grants, major contracts, industry cooperatives, centers for advanced technology, and university and industry consortia) and the range of major facility types for such linkages
(supercenters, incubator facilities, and established and new research
parks).

Colin H. Davidson gives the topic of incubator facilities fuller treatment
in Chapter 6. In discussing the incubator and regional and economic development, Section I of his chapter offers a transition between a planning
level and the specifics of particular projects. We can thus better place incubators in context, appreciate the variety of different types of incubators
and their role in local economic development, and understand the major

considerations to be addressed when planning incubators, together with financial and managerial aspects, and growth. Thus, far from being just another building type, successful incubators must address a mix of technological, managerial, and design considerations, including: a building, shared services, access to counseling on management and research and business-related matters and access to seed capital. This places this chapter's Section II in context, with its discussion of the architecture of the incubator, including design criteria at the feasibility, functional programming, and performance programming phases and procurement. Four projects are presented as examples.

## Part III: Managing and Designing Projects for High-Technology Workplaces

Included in this section are eight chapters examining various facets of the integration of technology, management, and design for productive work environments in high-technology workplaces. These chapters are grouped under three main subheadings: Human Purpose, Procurement Process, and Technological Means.

### HUMAN PURPOSE

An indirect impact of high technology has been that, with change as the only certainty, organizations have been undergoing unprecedented transitions. Facilities, as the name implies, should facilitate such transitions: Chapters 7 and 8 discuss aspects of organizations and facility management. A more direct impact of high technology has been the invasion of the workplace by a whole new range of electronic and telecommunication equipment, with numerous immediate environmental consequences for human needs, as discussed in Chapters 9 and 10.

Chapter 7, by Franklin D. Becker, explores the connections between facility management strategies and organizational culture. Its focus is university research laboratories, one of the critical facility types for high-technology activities, as seen in Ira Fink's Chapter 5. Research results emphasize the indirect role played by facilities in attracting and retaining high-quality faculty and students, and thus in strengthening the bargaining power—and ultimately the scientific vitality—of successful research units in competing for scarce resources. As a result of changes brought about by information technology, Becker suggests that facility management is becoming more proactive than reactive, and fully integrated with other planning functions. This implies research-based information, with systematic design and policy evaluation. Becker outlines an experimental approach to planning, designing, and managing university research facilities. He emphasizes the importance of maintaining a fit between facility management and organizational context, as well as the need to devote as many resources to this as to the creation of new organizational structures and management styles themselves.

This theme is echoed in the following Chapter 8 by Cecil W. Williams, who discusses the kernel fit between people, process, and place in high-technology organizations. He concurs with Becker in arguing that the workplace must become an integral part of management and of the business plan of an organization. Based on Jungian psychology and on the work of Kimberly and Quinn on the management of organizational transitions, his research on the "negotiable environment" has developed two models: the first gives us a grasp of different organizational cultures; and the second of individual work styles, with an analysis of corresponding characteristics of work processes and workplaces. Results demonstrate a lack of fit among people, work processes, and workplaces in many organizations. Thus, negotiating an improved fit is a major task for facility management.

T. J. Springer's Chapter 9, on the ergonomics of high-technology work environments, also supports similar views, but from the standpoint of anthropometrics and biomechanics, rather than from a social and organizational standpoint. He gives us an appreciation of the strategic role that ergonomics—its approach and methods—can play in managing the systemic fit among people, tasks, tools, and facilities. Today's workplace must accommodate both paper and electronic media, often with conflicting environmental requirements that may lead to a variety of misfits between people and their work environments. These may be mitigated, if not eliminated, with an ergonomic approach to problem solving. This approach is outlined by Springer and includes ways of defining the user population and of describing work with such tools as function analysis, function allocation, flow mapping, and task analysis as inputs to facility programming and design.

Springer's chapter clearly shows that environmental requirements for high-technology workplaces have become very complex, and that the accelerated pace of change is adding to this complexity. So, as Gerald Davis states in Chapter 10, "making buildings *work* is an extraordinarily complex task." To cope with the magnitude of this task, Davis outlines some current developments in standards setting for overall building performance, with particular emphasis on the impact of information technology. Beyond the measurement of individual performance criteria, such as those emerging from an ergonomic approach, these developments are intended to help the decision makers—managers, users, designers—in trade-off judgments, ratings, and the evaluation of *overall performance* (which is what really counts from the point of view of the occupants) to meet both functional needs and organizational requirements. This next evolutionary step in the historical progression of standards is discussed with reference to the ORBIT 2 study and the work of the ASTM Subcommittee EO 6.25 on Overall Performance of Buildings.

## PROCUREMENT PROCESS

Integrating technology, management, and design is a central preoccupation not only for programming, evaluating, and diagnosing facilities, as seen in the previous four chapters, but also for facility procurement. In Chapter 11, Colin H. Davidson reviews the options available in North America for the procurement of buildings for high-technology companies. For any new project, the building owner must bring together a task-oriented "temporary multi-organization." Davidson analyzes the sequence of procurement decisions and their underlying principles, and alternative options toward this end. For each alternative, this results in the coordination of a subset of a large number of potential players (through contractual links, agreed objectives, and modus operandi) despite the "seemingly impossible mix of interdependency and uncertainty" between them. Davidson discusses specific issues of procurement for high-tech facilities and advises the early consideration of procurement decisions for any new project.

## TECHNOLOGICAL MEANS

The last three chapters in this section focus on the technological means available to deliver high-technology workplaces: building technology and integrated information technology systems. A comprehensive review of technological means is offered by David B. Eakin in Chapter 12, with an account of the General Services Administration's Advanced Technology Buildings Program. Started in 1983, "this initiative emphasizes cost-effective innovation in all facets of the building design, providing enhanced building performance capabilities for the total work environment." In this light, facilities are seen as tools that contribute to the efficiency of the work process itself, and thus to the bottom line, responding to some of the user needs identified in earlier chapters. Major objectives include the need for flexibility, the reliability of services, the achievement of energy-conscious design, water conservation, low maintenance and servicing, and improved standards for fire safety and security. To such ends a major multifaceted program was established within the framework of a matrix where each functional issue is addressed by all building systems and features. These are reviewed under the headings of Architecture, Interior Design/Furniture, Structural, Mechanical, Electrical, Telecommunications, Fire Safety, Security, and Robotics. Specifically "high-technology" issues are thus fully integrated in this matrix, rather than considered independently. Here also the theme of the integration of technology, management, and design is underlined by the discussion of the design and construction practices intended to deliver such building features and to ensure that they operate as intended. Echoing Becker's call for an experimental approach to planning, designing, and managing facilities, Eakin promulgates the idea that our buildings should become laboratories "to teach us about our inventory and how it should be run."

In Chapter 13, Paul Stansall and Michael Bedford focus on a critical issue for all high-technology workplaces—that of cable management. Here again, the emphasis is on *management* rather than on the intricate details of the technology. Design flexibility to meet changing organizational needs is now largely conditioned by cable distribution. Thus, this chapter looks at how cables can be effectively managed physically, and what criteria must be considered jointly by managers and designers for cable distribution to contribute to organizational objectives and for buildings to become truly "intelligent." Cable management problems are reviewed, and two complementary approaches to problem solving are outlined: the kit of parts approach classifies building elements into shell, services, scenery, and sets, thus responding to different rates of obsolescence; while the traditional approach looks at buildings in orthographic projections to identify primary, secondary, and tertiary levels of distribution. A number of generic design solutions and solution types are then reviewed within this framework.

Fred A. Stitt's Chapter 14 discusses intelligent buildings further, as they are today and may become in the future given current trends in such areas as miniaturization, artificial intelligence (AI) and expert systems (ES), speech input/output, robotics, and artificial reality (AR). In the short term, centralized building systems management (CBSM) and shared tenant services (STS) are discussed. Stitt holds the view that all buildings, new or retrofitted old, will be "intelligent" by the year 2000. His speculations into "second-generation" futures challenges our imagination. While "old" future concepts, such as Dynabook and Xanadu, are becoming reality, some future possibilities, such as the miniaturization of computers down to the molecular level, have potentially far-reaching, if bewildering, consequences. They may contribute to raising the level of human consciousness, to reaching deeper for fundamental rationales for designing buildings—be they smart or dumb ones—and to bringing into focus our own primary standards of functional and aesthetic value.

## Part IV: Examples of Projects for High-Technology Workplaces

To conclude this book, two chapters have been assembled in this last section to illustrate projects for high-technology workplaces with specific examples. These have been selected to explore the reuse of old buildings for high technology, on the one hand; and a new high-technology greenfield development, on the other.

In Chapter 15, David Levy brings to our attention aspects of building reuse. As pointed out in the Preface, our intention is to look not only at the impact that the use of new technologies has on workplaces but also at their impact on light manufacturing production of high-technology hardware. The new production processes of high-technology companies will often be housed in retrofitted old industrial buildings that have become unsuited to their original purpose. A recent dramatic shift has occurred, away from the new and toward reuse, with 83% of total industrial, commercial, and institutional construction dollars spent in the United States for reuse projects during 1984, amounting to some $92.4 billion. Levy examines the major factors behind this shift and presents three reuse strategies: small buildings for small firms, large buildings made small, and large buildings for large firms. Incubator use often appears as a suitable candidate for redevelopment. Levy also discusses aspects of the feasibility of reuse projects, including the building/program match, codes, available technical knowledge about a vacant building, site location, vehicular access, regional and community site considerations, and cost estimating. He also warns us about important risk factors to watch out for. Finally, two examples serve to illustrate aspects of reuse projects.

In contrast to the reuse projects of Chapter 15, Chapter 16, by Harry A. Eggink and Robert J. Koester, presents a new high-technology project on a greenfield site: a 106-a prototype industrial setting in Indiana. Its program emphasized the role of communication and energy as principal design themes. The research program report resulted in a master plan and design guidelines, aiming to deal with both the specifics of this project and generalizable principles. It addressed three levels of issues, models, and alternative strategies (traditional, infrastructure, and energy park) and three scales of focus (site, building, space). Review policies and procedures were also carefully specified to ensure consistency in the development, including four stages in the review process for the design of individual buildings. A sense of quality of workplace for high technology emerges: "the sense of 'place' as a setting for human interaction."

# CHANGING ENVIRONMENTS

## High Environmental Quality for High-Technology Workplaces

This sense of quality is perhaps a key to the whole book: the high-technology workplace, as a shelter for knowledge-based activities, has brought about a qualitative shift in our perception of the workplace. Such activities can no longer be accommodated with minimal sheds and environmental conditions. It makes no sense to invest heavily in sophisticated hardware, software, and systems development, and yet to expect an effective use of such systems to occur—and to yield the anticipated returns—in old workplaces whose maintenance and upkeep, comparatively, have been starved of very basic investment over the years and that are somewhat reminiscent of Dickensian settings. In reuse projects, why consider retrofitting sophisticated "intelligence" when very basic comfort and code standards are hardly met? The implementation of work automation projects frequently brings into the spotlight some glaring inadequacies of existing work environments. These highlight the necessity to budget beyond automation per se to include such items as the workplace and training. And some managers might be dismayed to discover that such budget items can be a substantial percentage of total project cost, without an appreciation of the fact, argued elsewhere in this book, that such a "cost" is in fact an investment in productive environments.

Yet we are reminded of a fundamental consideration of interactive human/technology systems design, which is that no matter how sophisticated the design of the system, the "weak" link in the system is the person, and if people's needs are not catered for the system will operate only at a fraction of its expected efficiency. And, as mentioned previously, in competitive high-technology activities where attracting and retaining competent knowledge-workers are critical, such considerations are no "icing on the cake."

## Keeping Up with Technological Change

But investing wisely into a physical environment adapted to new technologies is a difficult task. Some potentially useful approaches and methods are discussed in this book. However, anticipating technological changes and corresponding time frames can be arduous, given the extraordinary diversity of a mercurial marketplace. Yet it has implications for judging the appropriateness of up-front capital investments in building systems with long depreciation periods. The commercial success of current cutting edge technical developments still in the laboratory could transform the hardware equipment and the workplace. A few examples may serve to illustrate this point.

While there is no doubt that wire management is causing considerable problems and costs (suspended floors are a must for most new "intelligent" buildings, space devoted to such floors, plant rooms, and control closets is increasing substantially at the expense of "inhabited" space), we could face a radical reappraisal if and when all types of transmissions (video, telephone, computer, integrated services digital networks [ISDN], local area networks [LANs], private branch exchange [PBX]) may access networks through universal ports and be merged into a common digital medium for routing.[17] While twisted pair is still the dominant medium, fiber-optic or digital FM transmission may soon (how soon?) dispose of a substantial proportion of cable management problems and render some current building features obsolete.

Similarly, another major problem has been the considerable cooling loads necessary for heating, ventilating, and air-conditioning (HVAC) systems in high-technology workplaces. These could be substantially reduced by recent developments: some optimists are looking forward to the day when we may use superconductive materials at ambient, rather than close to absolute zero, temperatures. And high-speed and energy-efficient gallium arsenide wafers may soon supplant silicone in a cost-effective manner.

Such developments also point to an imminent leap in miniaturization that together with flat screens (light-emitting diode [LED], plasma)—which may soon outperform traditional cathode ray tubes (CRTs) for equal cost—could lead to a dramatic reduction in the bulk of current desktop equipment, freeing horizontal work tops, causing the redesign of workstations, and having an impact on space planning. Indeed portable laptops might soon compete with PCs for comparable performance, and automated equipment might no longer be a quasi-permanent fixture of the workplace. Soon, such equipment may be no more obtrusive in the workplace than, say, the telephone has been for decades. Also, glare problems due to ambient lighting could be substantially reduced with the use of flat screens becoming widespread in the workplace.

Finally, the acoustic environment of the workplace may also be altered radically: noisy impact printers may soon be forgotten with the advent of competitively priced personal integrated laser printers/copiers. And voice input/output interfaces may soon demand as much attention as person/person voice communication does today.

## Controlling Changing Environments

Clearly it is difficult to fine-tune appropriate work environments for a variety of workers with different personal comfort thresholds carrying out a wide range of activities that change over time, and with the use of different pieces of equipment, each with different optimum environmental conditions. While centralized building systems management may take care of major comfort conditions "on automatic," we are now also witnessing a trend toward greater personal control and autonomy by occupants: Control over local temperature, ventilation, and noise levels. Control also over the suitability of the personal workspace for individual work styles and personalization.

Such decentralized environmental control also implies a participative style of environmental management (in keeping with the general trend toward the democratization of organizations), with people having a say in the initial planning of the workplace, and in subsequent changes.[18] If and when it becomes generally accepted that the physical environment affects job productivity and creativity—a point of view strongly supported by much of this book—we may soon witness *quality circles* in the workplace where the physical environment is routinely considered on a par with other factors such as managerial efficiency, systems and equipment, marketing and finances in order to improve overall organizational performance. Consensus seeking and/or conflict resolution techniques and procedures will also be needed at the interface between individual and group control. This also implies developing a sufficient degree of environmental competence among building occupants through training programs that address a range of issues from "how to adjust my chair" to "how can we use and adapt our facilities?"[19]

## Changing Organizations, Changing Environments

As organizations are affected by automation, space planning also is affected: both individual workstations and group spaces. Ellis has observed that automation may lead to more integration of functions, more group work, a differentiation of group types according to skill/task match, more lateral communication, more informal interaction, and more individualism.[20] Such organizational changes also have an impact on individual needs with an expansion of traditional job boundaries and a greater variety of functions that have different and sometimes conflicting environmental needs.

Therefore more space becomes allocated to meetings and group work functions, which frequently also include training functions. On the whole such space has hitherto remained relatively untouched by the new technologies that have affected individual workstations. A notable exception is videoconferencing facilities, which are not very prevalent yet. This may change in the not-too-distant future, as designers of information systems address the information needs of groups rather than individuals. Systems for *group computing* or *interpersonal computing* are in the development stage, such as the Colab system developed at Xerox's Palo Alto Research Center (PARC).[21] The equipment features a central electronic blackboard and interactive workstations for participants. Participants may also interact at a distance through LAN and PBX. Generally, *meetings technology* appears as a potentially fruitful, but as yet little developed, avenue of future development. This technology will have an impact on the "ecology of executive teams," a concept developed by Steele.[22]

As more space becomes devoted to either new technologies and related equipment or group work and training activities, space also becomes at a premium for individual workspaces. While overall office space per capita is on the increase, we can observe a trend toward the reduction of space standards for workstations and higher densities.[23] Indeed many furniture systems manufacturers make it a selling feature to fit more workers in less space. Less space does not necessarily imply reduced quality.[24] But, in relation to the considerable increase in the unit cost of usable—but "intelligent"—space, we should beware of minimum stress thresholds of "acceptable" crowding and territorial distances.

At the same time, there is a trend toward a reduction in the range of workstation sizes, toward "universal" sizes, in an effort to make these suitable to a wider range of worker functions: as workers can work in groups in designated areas as needed, as telecommunications reduce the need for physical proximity to information sources, and as audio and electronic surveillance can be used to supersede visual supervision, it is preferable, in so far as it is possible, to move workers around to suit organizational needs (rather than to reconfigure spaces and equipment) and thus to minimize the inevitable downtime and related cost penalties while also facilitating flexitime job and space sharing. However, the above trends run counter to some deeply ingrained attitudes in Western cultures about status and the personalization of the individual workspace: As observed by Wineman, "Aspects of the physical setting have symbolic status value in organizations. . . . They are perceived as part of the external reward system reflecting a worker's position and performance."[25] The conflict may be overcome by encouraging more "nomad" artifacts and symbols for the demarcation of status and personalization.

The above trends also run counter to many of the fundamental claims of increased social cohesiveness, communication, and productivity in open plan offices, as opposed to cellular layouts. Many such claims have already been substantially questioned, if not contradicted, by recent research,[26] and become undermined further by the impacts of IT. Rather than an either-or alternative between open plan and cellular layouts, intermediate models are being developed, such as the "cave and court" design suggested by Steele, with subtle hierarchical territorial transitions and levels of control between community and privacy.[27] Thus, workplace automation appears to magnify some issues about the appropriate degree of enclosure and visual and acoustic privacy, issues that must take into account functional and cultural factors specific to particular cases.

## Information Processing, Cultures, and High-Technology Workplaces

Therefore, for a definition of programmatic requirements for high-technology workplaces, cultural factors are just as important as strictly functional ones. Both Becker and Williams in this volume have observed the interdependence between facility management and organizational culture and management style.

But cultural differences, and how they may affect space perception and planning, may also be considered at various levels. Within a given country, for example, commonalties may be far more dominant than differences, and different organizations, while adopting apparently widely contrasting management styles, may still share many assumptions about appropriate work spaces—for instance, that privacy is highly desirable and conducive to improved performance or that physical status demarcation is essential to the social fabric of organizations.

The introduction of information technology affects the idiosyncratic information-processing styles of different cultures differently, hence also the flow of information in their social and physical work environments and their management styles. We may, for instance, characterize some cultures as *low context* (such as North American) and others as *high context* (such as Japanese).[28] Whereas in a low-context culture it is necessary to receive large amounts of specific briefing information as input to decisions, only a minimal amount is required in the high-context culture, where everyone constantly keeps abreast of contextual information. In the latter case, "channels are seldom overloaded because people stay in constant contact. Schedules and screening (as in the use of private offices) are avoided because they interfere with this vital contact."[29] Also, from-the-bottom-up group decision making by consensus and respect of established hierarchies are so ingrained in Japanese society that "the Japanese do not need status symbols such as private offices, or executive dining rooms to convey rank."[30]

A recent research report concurs:

*The level of office automation in the Japanese government is low, stemming from the management style practised. Except for the most senior managers there are no secretaries or private offices. Staff, middle management and even some people at the Director level work in offices containing 8–12 desks. Organization charts tend not to display authority relationships, but are a layout of the office, showing the location of each person's desk.[31]*

As most highly industrialized nations seek ways of beating the Japanese challenge, we may usefully ponder about such links between cultures, information-processing and management styles, information technology, and corresponding social and physical work environments.

## High-Technology Stakes and Changing Work Environments

As international competition on world markets intensifies and as ubiquitous information technologies transform our work environments, resources become more scarce and, as shown in this book, it makes sense to invest proportionally more into the physical environment of the workplace to improve its quality and shape it into a useful tool for management. But as the workplace increases its share of capital invested and contributes more to productivity and value added, so does it come increasingly to the forefront of issues of control and power in organizations.

When the information technology chips are down, the workplace moves toward the center of the bargaining table of labor relations. Stakeholders, apprehensive about the cloud of risk and uncertainty surrounding the introduction of IT, move to safeguard their interests. As pointed out by Ellis: "Space becomes a scarce resource for which different individuals and groups compete. Spatial allocation, subdivision and differentiation in terms of design and equipment are all likely to be affected by the structures of power within the organization."[32] As the structures of power vary from one organization to the next, so do management styles and attitudes toward the introduction of IT, together with its environmental component, with different degrees of controversy and conflict.[33] Taylorism and the subdivision of labor through scientific management have transformed blue-collar work and focused on a superficially efficient functional division of labor, while neglecting to take into account the effect of work conditions and the workplace on workers' morale and motivation, hence on productivity. Some argue that the introduction of IT into organizations may have the same effect on white-collar work unless we take steps to prevent it. Cooley's view, for instance, is that "the computer is the Trojan Horse with which Taylorism is going to be introduced into intellectual work,"[34] a move that is resisted in many countries by organized labor with concerns over alienating consequences and the deskilling of intellectual work. Cooley mentions a case where workers at the Norwegian Electricity Board opposed the proposed introduction of terminals "because they could only be operated in a mode which was 'unidirectional,' and hence not really responsive to the human being."[35] Some countries have given constitutional rights to workers in relation to IT:

*A recent Act in Norway requires employers to provide "sound contract conditions and meaningful occupation for the individual employee," "the individual employee's opportunity for self determination" and "each employer shall cooperate to provide a fully satisfactory working environment for all employees at the workplace."[36]*

It is worth noting that those concerns are not restricted to immediate physical health concerns, but also include the psychological well-being of workers.

## CONCLUDING REMARKS

As this book demonstrates, the introduction of IT leads to a much greater need for the integration of technology, management, and design. As aptly put by Williams in his Chapter 8, "The notion that the buildings should serve the *management* needs of corporations has not been widely promoted." As space becomes a tool of management, we need to be much better informed about its effectiveness. A considerable body of knowledge already exists that can contribute relevant information and methods, some of it already available in the form of guides.[37]

Yet as Shackel has noted with respect to ergonomics: "Although our knowledge is extensive, we should note the big discrepancy between what is known and how little of it has been applied in work situations."[38] He also makes a call to "develop a true synergy and symbiosis between ergonomics, computer professionals, architects and industrial designers," to which I would add others such as organizational ecologists, facility managers, and interior designers. We also need to devise and manage organizational mechanisms to ensure that the information that is available actually does reach decision makers in a timely fashion and in usable form and media. This implies an improvement in the level of environmental competence among stakeholders (including workers and management, but also designers) through training programs.

Also, we need to ensure that organizational mechanisms, tools, and methods are such that they facilitate incremental contributions to knowledge, from the point of view of the workplace both as a support system (building diagnostics and building science), as argued by Eakin in Chapter 12, and as a catalyst of social interaction and communication, as argued by Williams and Becker in Chapters 7 and 8.[39] While Eakin sees the workplace as an experimental laboratory, Becker puts forward an incremental "small wins" process that allows us to better grasp dynamic interactions in sociophysical systems. Environmental research may then better contribute to the process of appropriation of new technologies in organizations, with research informing practice, and vice versa, also adding to our understanding of the relationship between the physical environment of the workplace and its social and psychological environments.[40]

This represents a challenge of the highest order toward shaping the information-enhanced work environments of high-technology workplaces. The building where I work, at the Canadian Workplace Automation Research Center (CWARC), was inaugurated on November 5, 1985. Its inauguration plaque bears the words: "To translate technological progress into the humanization of the workplace: This is our challenge." This book contributes to that goal.

## NOTES

1. J. Skilling, "The Office Technology Task Force—Team Planning for the Integrated Electronic Office." *High Technology* (April 1983), pp. 69–82.
2. See John Worthington's Chapter 3 in this volume.

3. Herb Brody, "The High Tech Sweepstakes—States Vie for a Slice of the Pie." *High Technology* (January 1985), pp. 16–28. Also, see Charles Minshall's Chapter 4 in this volume.
4. See Lambert Gardiner's Chapter 2 in this volume.
5. See Worthington's Chapter 3 in this volume.
6. For example, "Linking the Factory and the Office." *High Technology* (November 1985), p.11.
7. CALUS. *Property and Information Technology: The Future for the Office Market* (Whiteknights, England: College of Estate Management), p. 24. Quoted in Walter B. Kleeman, Jr., "The Office of the Future," In Jean D. Wineman, ed., *Behavioral Issues in Office Design* (New York: Van Nostrand Reinhold, 1986), p. 279.
8. See Williams's Chapter 8 in this volume.
9. See Springer's Chapter 9 in this volume.
10. These figures are borrowed from Springer, in Chapter 8 of this volume. Other authors make the same point with slightly different figures. Wineman mentions, over a 40-yr life cycle of an office building, salaries will amount to 90 to 92%, with 2 to 3% spent on initial costs of the building and equipment, and 6 to 8 on maintenance and replacement. Jean D. Wineman, "Introduction—The Importance of Office Design to Organizational Effectiveness and Productivity," in Wineman, ed., *op. cit.*, p. xiii. Eakin mentions a figure of 91% for salaries in Chapter 12 of this volume, and states that if 10% is added to the cost of building while 1% is saved on employee efficiency, a return on investment of almost three to one can be realized.
11. See Springer's Chapter 9 in this volume.
12. M. Brill, S. T. Margulis, E. Konar, and BOSTI: *Using Office Design to Increase Productivity*, vol. 1., (Buffalo, N.Y.: Workplace Design and Productivity, 1984). pp. 55–69.
13. "Competitive High-Tech Market Turns Design Into Hiring Tool," *Contract:* March 1983, pp. 98–101.
14. See Becker's Chapter 7 in this volume in connection with university research labs.
15. See Minshall's Chapter 4 in this volume.
16. For the redesign of organizations in relation to the introduction of new computer systems, see Enid Mumford and Don Henshall, *A Participative Approach to Computer Systems Design—A Case Study of the Introduction of a New Computer System.* (London; Associated Business Press, 1979).
17. Dwight B. Davis, "Making Sense of the Telecommunications Circus." *High Technology*, September 1985, pp. 20–29.
18. Alan Hedge, "Open Versus Enclosed Workspaces: The Impact of Design on Employee Reactions to their Office." In Wineman, ed., *op. cit.*, p. 174.

19. A useful checklist of environmental education and training activities can be found in Fritz Steele, *Making and Managing High-Quality Workplaces—An Organizational Ecology.* (New York: Teachers College, Columbia University, 1986). pp. 63–65.

20. Peter Ellis, Presentation on "Workplace as Human Habitat" at Neocon 19, Chicago, Ill., June 11, 1987.

21. John Markoff, "Computing in Groups." *High Technology,* November 1986, pp. 56–57.

22. Steele, *op. cit.,* pp. 155–169.

23. CALUS, *op. cit.,* pp. 30, 32, 33, 39. In Wineman, ed., *op. cit.,* p. 280.

24. Anne Krueger, "Space Condensed and Improved in Citizen's Open Plan Redesign," *Contract,* March 1983, pp. 110–111.

25. Jean D. Wineman, "Introduction—The Importance of Office Design to Organizational Effectiveness and Productivity," in Wineman, ed., *op. cit.,* p. xv.

26. Hedge, *op. cit.,* p. 171. See also Brill et. al. *op. cit.,* pp. 94–103.

27. Steele, *op. cit.,* p. 98.

28. Edward T. Hall and Mildred Reed Hall. *Hidden Differences—Doing Business With the Japanese,* (Garden City, N.Y.: Anchor Press/ Doubleday, 1987).

29. Hall, *op. cit.,* p. 29.

30. Hall, *op. cit.,* p. 81.

31. Andrew Grindlay, *The Use and Management of Information Technology in the Government of Japan,* Canadian Workplace Automation Research Centre, Internal Report, January 31, 1987.

32. Peter Ellis, "Office Planning and Design: The Impact of Organizational Change Due to Advanced Information Technology," *Behaviour and Information Technology,* 1984, vol. 3, no. 3, pp. 221–233.

33. For alternative, and somewhat contrasting, postures toward the introduction of IT, see, for instance, the five contributions (by Michael J. Earl, Barry Sherman, APEX, Chris Harman, and Nuala Swords-Isherwood and Peter Senker) to Chapter 7: "Industrial Relations Implications," in *The Microelectronics Revolution—The Complete Guide to the New Technology and Its Impact on Society,* Tom Forester, ed. (Cambridge, Mass.: MIT Press, 1981. pp. 356–415.

34. Mike Cooley, *Architect or Bee? The Human/Technology Relationship.* (Boston, Mass.: South End Press; Hands and Brain Publications, Slough, England, 1982), p. 17.

35. Cooley, *op. cit.,* p. 41.

36. Act Relating to Worker Protection and Working Environment. Order No. 330, Statens Arbeidstilsyn Direktoratet, Oslo. Quoted in Cooley, *op. cit.,* p. 41.

**37.** See, for example, Arthur Rubin, *Interim Design Guidelines for Automated Offices*, 1984, revised 1986. (U.S. Department of Commerce, National Bureau of Standards, National Engineering Laboratory, Center for Building Technology, Building Physics Division, Gaithersburg, Md.) E. Cohen & A. Cohen, *Planning the Electronic Office.* (New York: McGraw-Hill Book Company, 1983). William L. Pulgram and Richard E. Stonis, *Designing the Automated Office—A guide for Architects, Interior Designers, Space Planners, and Facility Managers* (New York: Whitney Library of Design, 1984).

**38.** B. Schackel, "Information Technology—A Challenge to Ergonomics and Design," *Behaviour and Information Technology*, vol. 3, no. 4., 1984, pp. 263–275.

**39.** For the distinction of the environment as support and as catalyst, see Figure 1-2, and discussion in the text in Franklin D. Becker, *Workspace—Creating Environments in Organizations.* (New York: Praeger Publishers, 1981). p. 11.

**40.** Such as: Wineman, ed., *op. cit.;* Becker, *op. cit.;* Steele, *op. cit.;* and Eric Sundstrom and Mary Graehl Sundstrom: *Work Places—The Psychology of the Physical Environment in Offices and Factories* (Cambridge, Mass.: Cambridge University Press, Environment and Behavior Series, 1986).

PART ONE

---

# SOCIETAL TRENDS AND HIGH TECHNOLOGY

**2**

CHAPTER TWO

# FORECASTING, PLANNING, AND THE FUTURE OF THE INFORMATION SOCIETY

## *W. LAMBERT GARDINER*

## THE NEED FOR FORECASTING

### The Three Interfaces of Adam

Let me begin by providing you with a broad framework for what I am going to write. And just how broad is this framework? So broad that it will include not only everything that I am going to write but everything everyone else will write in this book, and, indeed, everything that anyone has ever written in the past and will ever write in the future.

The model can be called somewhat whimsically the Three Interfaces of Adam because it can be described in terms of the Christian cosmology. Imagine Adam all alone on our planet before it got so complicated. He had to deal only with the natural world—let us call it the *ecosphere*. Along came Eve and they prospered and multiplied, introducing another great sphere to Adam's environment, consisting of other people—let us call it the *sociosphere*. As Adam and Eve and their progeny made discoveries about and inventions from their environment, they built up a third great sphere, consisting of personmade things—let us call it the *technosphere*.

So here we have Adam or Eve, or you or me, in an environment represented by three spheres (Figure 2-1). The person, the only system within the universe that belongs to all three spheres, is in the center, which is the triple overlap of the three spheres. The person is the most complex system in the ecosphere, the natural world; the person is the element of the sociosphere, the social world; the person is the source of the technosphere, the artificial world. In order to distinguish between these three aspects of the environment, we can say that the ecosphere conforms to the laws of the natural sciences; the sociosphere, to the laws of the social sciences; the technosphere, to the laws of what Herbert Simon calls the "sciences of the artificial."[1]

## Scenarios for the Future

This framework was developed for the Ministry of Transportation and Communications of the Government of Ontario to synthesize a number of sources about the future that GAMMA (the future studies think tank in

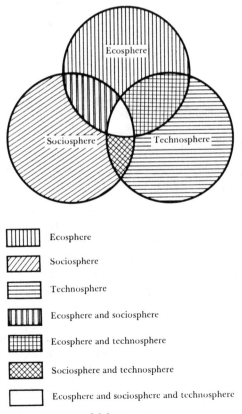

Ecosphere

Sociosphere

Technosphere

Ecosphere and sociosphere

Ecosphere and technosphere

Sociosphere and technosphere

Ecosphere and sociosphere and technosphere

**Figure 2-1.** The Three Interfaces of Adam.

Montreal to which I belong) and some other consultants had analyzed for them.[2] Did it work as a synthesis? Indeed, we think it did. Our future will be determined, as has been our past, by the complex interaction between these three vast spheres. Futurists differ, however, in the relative amount of emphasis they place on each of the three spheres.

Some argue that the ecosphere will become more important in the future than it has been in the past. Let us call this the *ecosphere-as-cause* scenario. Within this camp, there are pessimists and optimists. The pessimists, for example, the Club of Rome in their book *The Limits to Growth,* argue that we are going to destroy our civilization by using up our natural nonrenewable resources.[3] The optimists, for example, the GAMMA Group in our book *The Conserver Society,* argue that we can eke out those resources for considerably longer by conservation policies[4]

Some argue that the sociosphere will be relatively more important in the future. Let us call this the *sociosphere-as-cause* scenario. Once again, we have the pessimists and the optimists. The pessimists all the way from Thomas Malthus to Paul Ehrlich argue that the primary problem is one of overpopulation. The optimists are the advocates of capitalism and communism and—a third option for the Third World—the New International Economic Order, who argue that those people can be organized into productive systems that will generate the wealth to sustain them.[5]

Some argue that the technosphere will be relatively more important in the future. Let us call this the *technosphere-as-cause* scenario. Once again, we have the pessimists and the optimists. The optimists (for example, R. Buckminster Fuller in his book *Utopia or Oblivion*) argue that through technology each of us can live as kings lived in the past century[6] The pessimists (for example, Jacques Ellul in his book *The Technological Society*) argue that technology is not only not a solution but part of the problem.[7]

Within each pessimist camp, there is, of course, the no-future future. The ecosphere-as-cause doomsday scenario is that we will wipe ourselves out by using up our natural resources; the sociosphere-as-cause scenario is that we will destroy ourselves by overpopulating our planet; the technosphere-as-cause scenario is that we will blow our planet up with nuclear weapons. In "The Hollow Men," T.S. Eliot predicts: "This is the way the world ends/Not with a bang but a whimper." Doomsayers offer us a choice between two whimpers and a bang.

## Current Structural Shift

The current situation is characterized by a technosphere-as-cause scenario: the prime mover is information technology. Indeed, our future society has been defined in terms of this technology. We are, it is generally agreed, moving from an industrial society, based on energy, to a postindustrial society, based on information.

This is not simply a sectorial shift, within some subset of the technosphere, but a structural shift with reverberations throughout our entire model. It is important, then, that we see things whole. Hence the need for a broad model such as provided above.

You may reasonably say that this model is *too* broad. Indeed, it is. However, we academics tend to be too narrow in our thinking. If the person in the center is a natural scientist (physicist, biologist, etc.), he or she tends to look out over the ecosphere; if the person in the center is a social scientist (economist, political scientist, etc.), he or she tends to look out over the sociosphere; if the person in the center is an expert in the sciences of the artificial (architect, engineer, etc.), he or she tends to look out over the technosphere. They are like the three blind men, each in contact with a part of the elephant and getting a false view of the whole elephant.

You may reasonably say the model is too simple? Indeed it is. However, we academics tend to *under*-simplify things. It is a useful first slice of our complex modern reality. George Miller, in his classic article "The Magical Number Seven, Plus or Minus Two: Some Limits to Our Capacity for Processing Information" argued that we are capable of handling simultaneously only seven categories (give or take a couple).[8] The model has the optimal number of seven categories.

This broad and simple model permits us to digest our total situation in one "eye gulp." It allows us, like an artist working on a large canvas, to stand back from time to time to see the whole painting. Now, let us move in to work on some of the detail.


## THE NEED FOR PLANNING

### From One- to Two-Story Model

During these transitional turbulent times, it is important that we, as individuals and as representatives of our various institutions, plan for the emerging new society. One way is to sensitize ourselves to the trends from the past and extrapolate them into the future to get some sense of where we are going. The study of trends has become a trend in itself, which could, with apologies to John Naisbitt, be called a metatrend.[9]

However, it is not enough simply to watch the trends—one must do something with this "intelligence" about our environment. As we move from forecasting to planning, we must move to a more elaborate model. The different scenarios for the future described above within the Three Interfaces of Adam model—ecosphere-as-cause, sociosphere-as-cause, and technosphere-as-cause—suggest that our future will be determined by our various environments. It is necessary for us to go beyond such deterministic scenarios to consider the person-as-cause scenario.

One encouraging sign of the recent times is that we are beginning to step back and look at more of the elephant. Some people are looking at the overlap of ecosphere and sociosphere (for example, studies of the effect of legislation about conservation on our natural environment); some people are looking at the overlap of sociosphere and technosphere (for example, social assessment studies of the human impact of technology); some people are looking at the overlap of the ecosphere and the technosphere (for example, technology assessment studies of the environmental impact of technology).

We must, however, step even further back and look at the *triple* overlap of the three spheres—the person in the center. Our future will be largely determined by the actions of people, whether through the mediation of the ecosphere, the sociosphere, or the technosphere or whatever complex combination of our different environments.

We must look not only *at* the person in the center but also *within* the person in the center. Social change is determined by the actions of the person in the center, but those present actions are based, in turn, on values acquired in the past and visions of the future. It is necessary, therefore, to erect a second story in our model, which represents the subjective map of the person in the objective world (Figure 2-2).

## The Prospective Method

Our future will be determined not only by trends from the past but also by visions of the future. Michel Godet, in his book *The Crisis in Forecasting and the Emergence of the Prospective Approach,* argues for the need to consider visions as well as trends.[10]

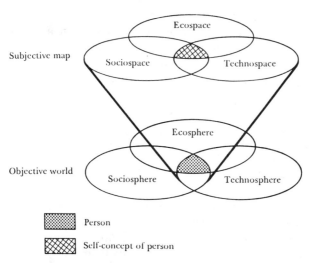

**Figure 2-2.** Subjective map of objective world.

The *prospective approach,* considers the future "not as something which is already decided and which gradually reveals itself to us but as something which is to be created."[11] In using this approach "one begins by defining ends which are noble enough to be generally pursued and thus can be incorporated into the culture of the society. Then action can be defined through a constant interplay of ends, available means, and present reality."[12] This method has been advocated and used by Kenneth Boulding, Bertrand de Jouvenel, Willis Harman, and many other scholars.[13]

The prospective approach makes the commonsense assumption that in order to get from A to B, one must first know where A is and where B is. In this case, A is the communication system before the current wave of technological innovations and B is the communication system after the revolution. The shift from trends to visions requires "imagination and the ability to conceive of utopias."[14] Our values can be embedded in our visions of a desirable future. Unless we make the shift from trends to visions, our future will be determined by facts rather than values. To quote the famous philosopher Casey Stengel, "If you don't know where you're going, you're likely to end up someplace else."

## Visions of the Information Society

The two major visions of the emerging information society, according to my colleague Kimon Valaskakis, president of the GAMMA Institute, are the *telematique* scenario and the *privatique* scenario.[15] The telematique scenario, based on television, envisions a few large sources beaming information down a great electronic highway to millions of destinations. The privatique scenario, based on telephony, envisions a network of nodes, at which everyone is source as well as destination.

As a psychologist, my role is to nag him about this neat dichotomy being complicated somewhat by the "pique" scenario, in which people say, "They are building better and better mousetraps and selling them to the mice—it is time to squeak up." What will eventually emerge is some complex amalgam of these scenarios.

## THE NEED FOR CONVIVIAL ENVIRONMENTS

### Pessimistic Vision of Physical Environments

In *The Three Boxes of Life,* Richard Bolles argues that we typically lead our lives in three boxes: the education box, in which we spend most of our time learning; the work box, in which we spend most of our time working; and the retirement box, in which we spend most of our time playing.[16] We tend to have three sets of space boxes to correspond to those three sets of time boxes. We learn in the school box, work in the office/factory box, and play in the home box. Let us look at the pessimistic vision of the emerging information society, with respect to each of these "boxes" in turn.

The pessimistic vision of the school of the future is simply that it will be a continuation of the school of the present. Our philosophy of education is embodied as much in mortar as in mortar boards: rows of desks bolted to the floor facing the "front" of the classroom, where stands the teacher, the source of all knowledge. This traditional "outside-in" image of teaching made some sense in the early days of universal education, when the personal knowledge of the teacher, limited as it must be, could somewhat enrich information-impoverished environments. However, it makes no sense now that our everyday environment is so rich in information that a child's education is interrupted by going to school.

The pessimistic vision of the office/factory of the future is based on there being a repetition of the errors of the first wave of automation. When blue-collar work was automated, time-and-motion studies attempted to integrate the person with the machine to maximize the efficiency of the process of production. But the limitations of this approach were revealed in an experiment at the Hawthorne Works of the Western Electric Company in the 1920's, in which researchers were investigating the effect of working conditions on the productivity of workers.[17] One variable was illumination. They increased the illumination. Production went up. They decreased the illumination. Production went up. They pretended to change the illumination but actually left it the same. Production went up. The same surprising results were found when they varied a number of conditions in the physical environment of the worker. Production could be increased by doing something—indeed, anything. It did not really matter precisely *what* was done.

This Hawthorne effect has been attributed to the fact that any novelty decreases boredom and thus increases productivity, or to the fact that the researchers' interest in the workers increased morale and thus productivity. Emphasis in industrial psychology shifted, as a result of this experiment, from the physical environment of the workplace to its psychological climate.

In this second wave of automation—the automation of *white-collar* work—many people have ignored the lesson of the Hawthorne effect and have repeated the error of time-and-motion studies. They assume that productivity can be maximized by manipulating the physical environment. Secretaries are corralled into word-processing pools (or slightly less centralized puddles), organized in anonymous lanes of workstations, and monitored keystroke by keystroke by feedback from the word processor.

This office-as-factory model is also seen in the "electronic sweatshops" springing up in suburban regions.[18] Corporations (Bank of America, Pacific Telephones, AT&T, Mobil Oil, and others) are building cheap back offices in the suburbs to house their rote work. Low-paid workers sit at rows and rows of cathode ray tubes in these word- and number-processing factories.

The pessimistic vision of the home of the post-industrial society is embodied in enclave theory. This is the argument that the home of the future will be a sort of "womb with a view" (or, for those more sociable, a womb for two, or perhaps a few, with a view). The view will be provided by a television screen. This screen will be more a window on the world outside than a mirror of the world inside as it often is now. Through this window will pour information about the world outside, through a multitude of channels over various delivery systems. However, the window will be two-way. The authorities outside will be able to monitor the occupants inside. Teleshopping and telebanking services through the screen will further reduce the need to ever leave the womb. The home will become a fortress to protect its occupants from an increasingly hostile environment.

E. M. Forster takes this enclave theory vision to a terrifying extreme in his short story "The Machine Stops."[19] In his vision people live in individual capsules, which provide satisfiers of all their basic human needs. They are, thus, totally dependent on the machine to which the capsules are attached. And one day, the machine stops.

## Optimistic Vision of Physical Environments

The above-mentioned book *The Three Boxes of Life* has a subtitle—*And How to Get out of Them*. After documenting the difficulties of living within each of the three boxes and of moving easily from the education box to the work box and from the work box to the retirement box, Bolles suggests how we may get out of them (Figure 2-3). In the few years since he wrote the book, electronic technology has developed to a point at which it may be able to help him in his project of prying us out of those boxes.

Let us turn now to the more optimistic scenario, in which the potential of electronic technology to help us lead more integrated lives is realized. How can our new information-processing tools be used to pry us out of our three life boxes and, more relevant to this context, out of the three physical boxes in which the specialized activities of learning, working, and playing are located?

In the information society, our three "boxes" could be strongly influenced by the transportation/telecommunications trade-off. An anecdote may be helpful in introducing this topic. In the early 1970s a friend of mine visited Marshall McLuhan. Over lunch, McLuhan said casually, "Executives drive to the office to answer the telephone." When George passed on this off-the-cuff remark when he got back, he triggered a 6-hour discussion. We all knew that many business people spend many gas-guzzling, air-

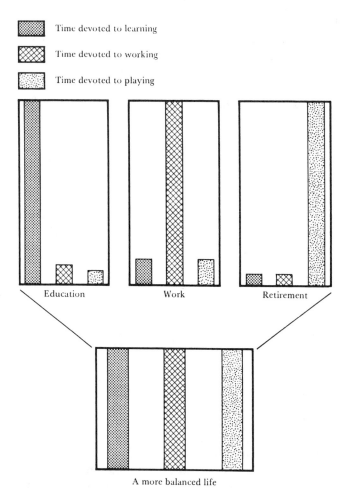

**Figure 2-3.** Getting out of the Three Boxes of Life (Bolles).

polluting, time-wasting, ulcer-creating hours commuting and that they spend many of their office hours on the telephone. However, it took the genius of McLuhan to put those two facts together and point out that they had a perfectly good telephone at home.

Since then, every organization whose mandate has anything to do with either transportation or telecommunications seems to have conducted a study on the transportation/telecommunications trade-off.[20] They all see it as a challenge—the former group as a threat/challenge and the latter group as an opportunity/challenge.

It is an idea whose time has perhaps come. Now that the telephone receiver is being supplanted by terminals, the telephone system can be used to communicate with computers as well as with people. It has been predicted that such telephone/terminals will outnumber traditional handsets by the year 2000.[21] Computers are lousy conversationalists but are very well informed. People who need information to perform their work can now collect much of that information by "letting their fingers do the walking." It is therefore now possible to "compute" to work. The modern interpretation of the transportation/telecommunications trade-off is "sell your car and buy a computer."

It is important to distinguish between delocalization (the distribution of places and, hence, people) and decentralization (the distribution of power).[22] The transportation/telecommunications trade-off permits delocalization but not necessarily decentralization. Indeed, the evidence suggests that telecommunications contributes to centralization. Captains of ships and ambassadors used to have considerable autonomy. Telecommunication permits wars to be waged by the admirals on shore and peace to be negotiated between leaders in capitals, with the people in the field merely following orders from headquarters.

With vast sources of information available literally at one's fingertips, teachers are relieved of the traditional outside-in information-providing function of teaching. This sets them free for the more human inside-out inspiration-creating function. Computers can provide only data. Teachers are required to put this data in context to yield information, information in context to yield knowledge, knowledge in context to yield understanding, and understanding in context to yield wisdom. The school of the future will be characterized by this synergistic relationship between computers and teachers to help people up this hierarchy of content within context from data to wisdom.

Maison Alcan, the new headquarters of the Alcan Corporation in Montreal, could be a model for the corporate office of the future. Because of the transportation/telecommunications trade-off, it is no longer necessary for all the main office employees to be huddled under one roof, as were the Alcan employees in their old headquarters in the high-rise Place Ville-Marie. They were thus able to build a small, high-prestige set of offices as a headquarters hub, with other employees delocalized to the periphery.

The office of the architect and the engineer will be transformed by our new electronic colleagues. A few years ago, I was awed on visiting the New York Institute of Technology to see a mathematician create a three-dimensional image on a screen from formulas and explore the three-dimensional space he had created by "flying" through it using a joystick as in a plane. That, however, was the most sophisticated computer-aided design system available at the time, worth millions of dollars. Recently, Don Collins, a colleague at Concordia University, who had resigned to found ACADZ (that is, Computer-Aided Design from A to Z), showed me essentially the same system on an IBM PC.

In many cases, the home has degenerated into a sort of service station, where we fill up during the day and park at night, after spending most of our time at the office/factory or school. Electronic technology permits us to work and learn at home. The home of the future could be the place where we escape from the three boxes of life into a more integrated life in which we flow freely between learning and working and playing.

People are beginning to escape from the rigid set that work is done in an office/factory (the two words are collapsed since they are becoming more and more difficult to distinguish) and learning is done in a school/university. I once overheard one editor at my publishing company saying to another: "I won't be in the office tomorrow, Terry, I've got some work to do." Recently, a current colleague of mine blurted out, when asked about his schedule: "I work in the morning and come to the office in the afternoon."

Attitudes of the other members of a family will also have to change to make "homework" feasible. A professor on sabbatical, trying to write a book at home, was constantly interrupted by his wife and children. They had to establish a convention. After breakfast together, he would put on his coat, pick up his briefcase, kiss them all good-bye, and walk out the front door and in the back door to his office. As far as everyone was concerned, Daddy was at the office until the ritual was repeated in reverse in the evening.

The physical layout of the home will also have to change. Interesting problems are posed to architects to design an appropriate mix of public and private spaces. The typical home, consisting of living room, kitchen, bathroom, and x bedrooms, betrays the current emphasis on basic biological functions of eating and sleeping. The home in the information society will contain offices for working and entertainment/education centers for learning. Bedrooms may become smaller to make more space for learning, working, and playing.

## Threats or Opportunities?

Whether this paradigmatic shift from an industrial to an information society is viewed as a threat or an opportunity is largely a matter of attitude. To use the language above, the same process in the objective world can be either a threat or an opportunity in our various subjective maps of the objective world. A recent book entitled *Crises are Opportunities* has a graphic on the cover containing the almost identical Chinese characters for the words "crisis" and "opportunity."[23]

Whether the pessimistic or the optimistic visions of our physical environments in the information society will prevail is, once again, largely a function of our various subjective maps. Engineers and architects, who are responsible for creating convivial physical environments for human activities like learning, working, and playing, can influence the information society with their negative or positive visions of that society. Those visions will, consciously or unconsciously, be built into the places where we learn, work, and play in the future.

## NOTES

1. Herbert A. Simon, *Sciences of the Artificial* (Cambridge, Mass.: MIT Press, 1969).
2. GAMMA Group. *"Tracking" Synthesis: Report on the Future of Ontario*, commissioned by MTC, Ontario Government (Montreal: GAMMA, 1981).
3. Donnell H. Meadows, et al., *Limits to Growth* (New York: Universe, 1974).
4. K. Valaskakis, et al., *The Conserver Society* (Toronto: Fitzhenry & Whiteside, 1979).
5. Documentation of the capitalist and communist options for the first and second worlds is very familiar. The third option for the Third World is not quite as familiar—here, then, are two sources: Revin Laszlo and Joel Kurtzman, *The United States, Canada, and the New International Economic Order* (New York: Pergamon Press, 1979). J. A. Tinbergen, J. Dolman, and J. Van Ettinger, *Reshaping the International Order: A Report to the Club of Rome* (New York: E. P. Dutton, 1976). The pessimistic version of the sociosphere-as-cause scenario was updated in 1980, when three widely divergent sources cited population as the major problem of the globe: Willy Brandt, *North-South: A Program for Survival* (Saint-Amand France: Editions Gallimard, 1980). Alexander King, *The State of the Planet* (Willowdale, Ontario: Pergamon Press, 1980). Gerald O. Barney, study director, *The Global 2000 Report to the President: Entering the 21st Century* (Washington, D.C.: U. S. Government Printing Office, 1980).
6. R. Buckminster Fuller, *Utopia or Oblivion* (Carbondale, Ill.: Southern Illinois University Press, 1969).
7. Jacques Ellul, *The Technological Society* (New York: Random House, 1964).
8. George A. Miller, "The Magical Number Seven, Plus or Minus Two: Some Limits on Our Capacity for Processing Information," *Psychological Review*, 1956: 63, pp. 81–97.
9. John Naisbitt, *Megatrends: Ten New Directions Transforming our Lives* (New York: Warner, 1982).
10. Michel Godet, *The Crisis in Forecasting and the Emergence of the "Prospective" Approach* (New York: Pergamon, 1977).

11. G. Berger, "The Prospective Attitude," in *Shaping the Future*, A. Cournand and M. Lévy eds. (New York: Gordon & Breach Science Publishers, 1973), pp. 245–249.
12. B. J. Huber, "Images of the Future," in *Handbook of Futures Research*, J. Fowles ed. (Westport, Conn. Greenwood, 1978), pp. 179–224.
13. Kenneth E. Boulding, *The Meaning of the Twentieth Century* (New York: Harper & Row, 1965). Bertrand de Jouvenel, "Introduction," in *Futuribles: Studies in Conjecture*, B. de Jouvenel, ed. (Geneva: Droz, 1963), pp. ix–xi. Willis H. Harman, "On Normative Forecasting," in *The Study of the Future: An Agenda for Research*, Wayne J. Boucher, ed. (Washington, D.C.: National Science Foundation, 1977).
14. Langdon Winner, "On Criticizing Technology," in *Technology and Man's Future*, 2nd ed., Albert H. Teich, ed. (New York: St. Martin's Press, 1977).
15. Kimon Valaskakis, *The Information Society: The Issues and the Choice*, Information Society Programme: Integrating Volume (Montreal: GAMMA, 1979).
16. Richard N. Bolles, *The Three Boxes of Life: And How to Get out of Them* (Berkeley, California: Ten Speed Press, 1981).
17. F.J. Roethlisberger and W.J. Dickson, *Management and the Worker* (Cambridge, Mass.: Harvard University Press, 1939).
18. Peter Calthorpe, "The Back Office: Post-industrial factories," *Whole Earth Review* (January 1985): 44, pp. 30–31.
19. E.M. Forster, "The Machine Stops," in *The Eternal Moment and Other Stories* (New York: Harcourt Brace Jovanovich, 1956).
20. For example: A.J. Cordell and J. Stinson, *Travel and Telecommunications: Survey Results to Date and Future Possibilities* (Ottawa: Science Council of Canada, November 1979). E.S. Darwin, "The Potential of Telecommunication Innovation on Intercity Passenger Transportation in Canada," *Transportation and Telecommunications*, vol. 1 (Ottawa: Strategic Planning, Transport Canada, January 1982). Samuel W. Fordyce, *NASA Experience in Telecommunications as a Substitute for Transportation* (Washington, D.C.: NASA Headquarters, April 1974). A.H. Kahn, *Transportation and Telecommunications: A Study of Substitution and Their Implications*, report no. 121, 1706 (Hull: Canadian Transport Commission, June 1974).
21. Peter Marsh, *The Silicon Chip Book*, (London: Abacus, 1981).
22. Mario R. Espejo and Jean-Claude Ziv, "Communication, Delocalization of Work and Everyday Life," in *Information Technology: Impact on the Way of Life*, Liam Bannon, Ursula Barry and Olav Holst eds. (Dublin: Tycooly, 1982), pp. 215–232.
23. Michel Godet, *Crises are Opportunities* (Montreal: GAMMA Institute Press, 1985).

# 3

## CHAPTER THREE

# ACCOMMODATING THE KNOWLEDGE-BASED INDUSTRIES

## *JOHN WORTHINGTON*

In this chapter I will attempt to identify the nature of the emerging high-technology companies and their requirements for premises by focusing on examples of firms and developments in Europe. I will cover three main themes: firstly, a review of the changing pattern of employment and the impact of changing technologies; secondly, a review of buildings and sites currently being developed in Europe to accommodate the needs of emerging industries; and finally, examples of how specific firms are using their space.

Historically, work was undertaken in two distinct environments: the office and the factory. Today, these definitions have become blurred. The point can be exemplified by two contrasting images. The first is of a row of ladies sitting at work tables making Mickey Mouse telephones. They have a carpeted floor, high light levels, a clean, cheerful environment, and working densities similar to those in any clerical office environment. The second image is again of women sitting at desks, each with a visual display unit; there is carpet on the floor and a clean environment, and they are organized in a production line layout to process data. In environmental terms it is difficult to distinguish between the office and the factory, or the factory and the office.

The disintegration of traditional office uses has become most apparent in the electronics field where in a customer-support branch office, one will find product repair, customer services, training, and demonstration, as well as more traditional marketing and administrative office functions. I will now draw on three distinct threads of the experience of my architectural firm, DEGW, in London, in

- advising multinational electronics companies in planning and fitting out space, much of which was for nonoffice functions;

- adapting old redundant industrial buildings in inner city locations for multiple use by small emerging firms;

- planning new developments on greenfield sites to match the needs of emerging modern industry.

## CHANGING PATTERN OF EMPLOYMENT

In the past 50 years the pattern of employment in the United States has seen a dramatic shift from agriculture and manufacturing, where the labor force has declined, to the service and information-processing sectors, where the labor force has greatly increased (Figure 3-1). In the

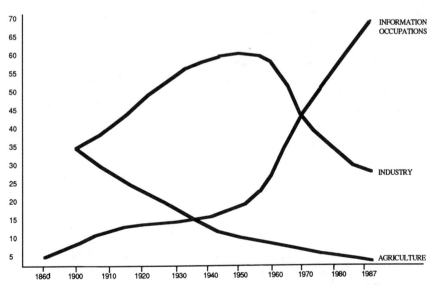

PERCENT OF U.S LABOUR FORCE

**Figure 3-1.** Changing pattern of U.S. labor 1860-1980 (source: Parker and Porat, 1975, with additional information to 1987 from Employment Gazette, Department of Employment, U.K.).

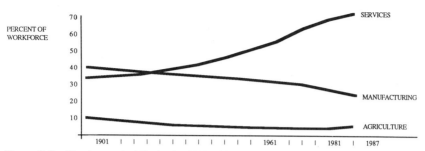

**Figure 3-2.** Employment shifts in the U.K. 1901-1987 (source: Employment Gazette, Department of Employment, U.K.).

United Kingdom we are experiencing the same trends but somewhat later (Figure 3-2), with a major increase in the numbers employed in the service and office sectors. This is currently reflected in our educational system, where youngsters are being trained in clerical and typing skills to go into insurance and banking. As the office function automates, will employment demand also decline in these sectors as it did in shipbuilding in the past? In the United States between 1982 and 1984 roughly 165,000 jobs were lost in steel making, while the same number were created in the production of computer hardware. We are faced with a restructuring of work: from the use of brawn power to brain power, and away from the mass production of products to the application of ideas. Figure 3-3 reflects the change to the application of knowledge. In the 1930s the components of a product were high labor content, lots of materials, some energy, and small amounts of finance and information.

The big change in the 1980s has been the application of information, less labor, and a lot more expenditure of energy. The skills required are

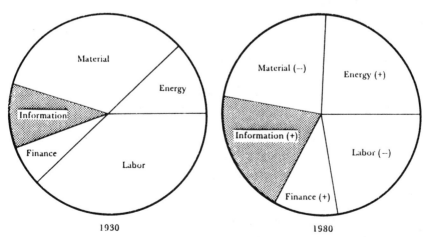

**Figure 3-3.** Knowledge is increasingly important as a factor of production (source: Daniel Bell: "Post-Industrial Society").

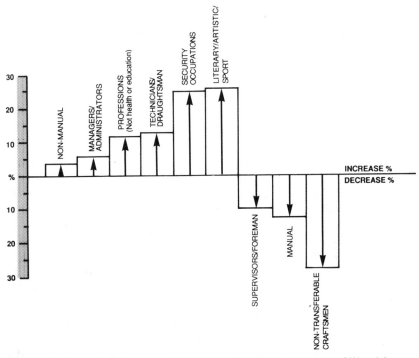

**Figure 3-4.** How jobs will rise and fall 1980-1990 (source: University of Warwick Institute of Employment Research).

shifting from supervisory, manual, and nontransferable crafts to managerial, professional, and technical (Figure 3-4). The areas where jobs are increasingly available are security, janitorial, and cleaning functions, the arts, and part-time work.

## CHANGING TECHNOLOGIES

The microchip is the core of a widening range of products. We have seen a shift from electromechanical to electronic products, and within the electronic sector in the past 20 yr there has been a shift from mainframe computers that required whole rooms to desktop micros with the same amount of power. The microchip is becoming smaller, more powerful, and cheaper, with increasing applications. The result is a converging of technologies into single products that perform several different functions (Figure 3-5). Where recently there were separate systems for storing, processing, and communicating information in the office, these are now converging into single systems.

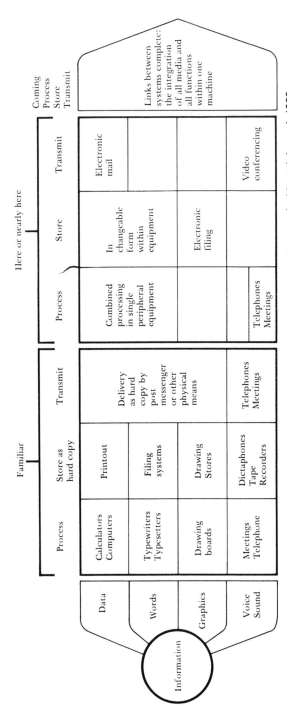

**Figure 3-5.** Information forms and processes are converging into multi functional products (source: Architects' Journal, 1982; John Worthington, senior partner, DEGW Architects, Planners, and Designers, London, Glasgow, Paris, Milan).

## CHANGING METHODS OF PRODUCTION

The speed of innovation and the flexibility of production have resulted in shorter product life spans. The average life span of a product is 18 months, and it is only the most successful products that pass from the pilot production line to mass assembly. The goal is to maximize on the latest and best ideas of technologies and to minimize production costs, and we are seeing the emergence of global products—product development in America, software development in Europe, and the chip mass-produced in Taiwan, with assembly and customization in the country of sale. A few mass-produced components can be assembled into a range of customized products for precise markets.

## CHANGING PATTERNS OF WORK

Short product life cycles reflect a shift away from mass assembly to small-batch production where jobs may change every few hours, and teams are formed in which there is a strong overlap between the engineering, design, and manufacturing process. Learning curves may be reduced by interactive video. Processes even for short runs are susceptible to automation, with links between computer-aided design (CAD) and manufacturing processes.

As repetitive jobs become automated and information transference from computer to computer improves, the organization of work becomes more fluid. Instead of being tied to one workplace, workers in an office or factory now move between workplaces—from their own desks, to a CAD terminal, to a meeting room, to a project area, and back to their desks. Others may with "laptop computers" and modems decide to spend part of the day working from home, at a hotel, or even on a train. Flexitime, job sharing, and "telecommuting" are all viable ways of organizing production.

## "HIGH-TECHNOLOGY" ENVIRONMENT

In discussing the environments required by high-technology companies, we need to differentiate between the product produced and the manufacturing process used (Figure 3-6). The craftsperson in the garage "producing a wooden toy" is producing a low-technology product with a low-technology process. At the other extreme a microchip is a high-technology product with a high-technology manufacturing process. A

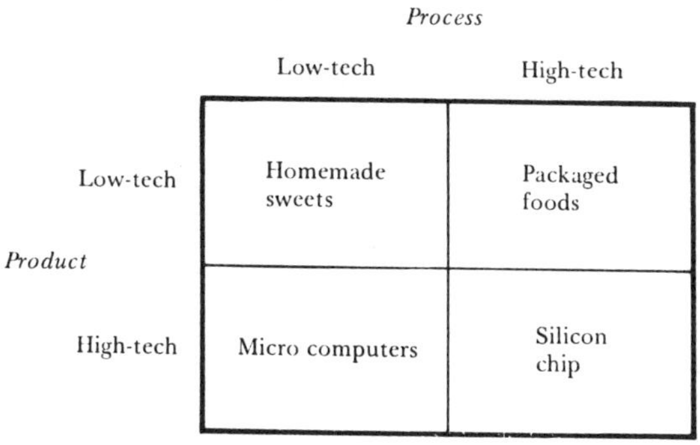

Process

Low-tech          High-tech

| | Low-tech | High-tech |
|---|---|---|
| Low-tech | Homemade sweets | Packaged foods |
| High-tech | Micro computers | Silicon chip |

Product

**Figure 3-6.** High technology refers to both product and processes.

microcomputer, however, is a high-technology product with a low-technology process.

The majority of the high-technology products are low-technology assemblies of highly sophisticated components that are produced in high-technology plants. Many of these high-technology products with low-technology production processes are assemblies of components to meet specific market demands, produced as one-off items or small batches. To tailor such products requires rapid reaction to market needs and the new technologies available.

It is natural then that in a building producing telecommunication systems we should find under one roof the salespeople, who are the antennae to the users' requirements, sitting next to the research and engineering design group, who apply the latest knowledge and design the product. These in turn are intimately linked with the component assembly and testing area where the final system is produced, and finally there are administrative and accounting staff. All these functions come together to allow the latest knowledge to be applied in providing a rapid response to the marketplace. Figure 3-7 represents the range of functions that traditionally were to be found in distinct locations, building types, and environments and are now merging.

Firms, like families, follow a life cycle and at each stage have different accommodation needs. In their infant stage they may rent space in a larger building and merge the functions of corporate headquarters, research, pilot manufacturing, and marketing in one place. As the organization grows, functions become more specialized until the firm becomes institutionalized with established products and probably has separate locations for a research campus, a downtown corporate headquarters, manufacturing plants, and local marketing branches and customer service centers (Figure 3-8).

| | Production | | | Support | | |
|---|---|---|---|---|---|---|
| | Research and development and customization | Distribution and storage | Assembly and servicing | Customer support | Sales and marketing | Administration |
| Firm A | | ✔ | ✔ | ✔ | | |
| Firm B | | ✔ | ✔ | ✔ | ✔ | ✔ |
| Firm C | ✔ | ✔ | ✔ | ✔ | | |
| Firm D | ✔ | ✔ | | | ✔ | ✔ |
| Firm E | ✔ | | | ✔ | ✔ | ✔ |
| Firm F | | ✔ | | | ✔ | ✔ |
| Firm G | | ✔ | ✔ | | | |
| Firm H | ✔ | ✔ | ✔ | ✔ | ✔ | ✔ |
| Firm I | ✔ | | | ✔ | | ✔ |
| Firm J | | | | ✔ | ✔ | ✔ |
| Firm K | ✔ | | | | | ✔ |
| Firm L | ✔ | | ✔ | | | |

**Figure 3-7.** Range of functions often now found in one building.

What sorts of developments will these different types of organizations require? An analysis of developments DEGW undertook in the United Kingdom and Europe (Table 3-1) identified five distinct types of development for the knowledge-based industries:

1. Innovation Centers.
   Multitenanted light industrial buildings, immediately adjacent to or on a university campus, providing small units (100 to 500 sq ft, or 9.3 to 46.5 sq m) for firms growing out of existing research groups within the university.

2. Science/Research Parks.
   Developments aimed at growing or established firms in research or development. Associated with a university campus, and jointly sharing amenities, research facilities, and know-how. Sites of 5 a or more. The establishment of "corporate research campuses."

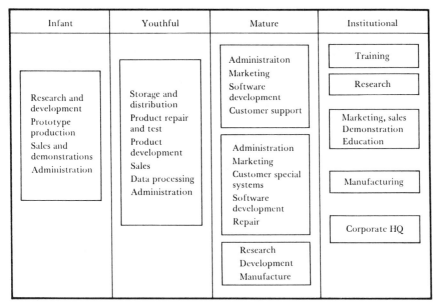

**Figure 3-8.** Functions specialize into individual buildings as the organization matures.

**3.** Technology Parks.

Commercial developments, in a location with good academic institutions, and an attractive life-style to appeal to scientific and professional staff. Low-density well-landscaped development of 100 a or more.

**4.** Business Parks.

High-quality, low-density environment associated with excellent road and air communications and a metropolitan catchment area. Aimed at firms requiring a prestige image and high-caliber work force. Mixture of manufacturing, product assembly and customization, and sales and training functions.

**5.** Upgraded Industrial Estates.

Many of the so-called European high-technology developments are an upgraded marketing image applied to a standard industrial estate, where the industrial warehouse sheds have had a "high-tech" facade applied.

As we begin to learn more about the demands of emerging industries, it is becoming clear that there is not one solution. Different stages in the life cycle require different solutions for premises, and the aspirations of the firm, depending on whether it is craft oriented or entrepreneurially inclined, will affect the quality and character of the buildings (Figure 3-9).

**TABLE 3-1.**
**Range of developments for new knowledge-based growth industries**

| | | |
|---|---|---|
| INNOVATION CENTER | Industrial building. Immediately adjacent to university campus, providing small units for firms growing out of research projects or expertise within the university. Existing building. | Enterprise Lancaster |
| RESEARCH PARK | Development aimed at growing or established firms in research or development. Associated with university research laboratories and amenities. Workshop, laboratory, office functions. Joint venture with or supported by university. | Heriott Watts Cambridge Science Park Leuven, Belgium |
| SCIENCE AND TECHNOLOGY PARK | Universities and research institutions with 30 mile radius. Attractive life style to appeal to scientific and professional staff. Low-density development. | Birchwood, Warrington Kirkton Campus Livingston |
| COMMERCIAL/ BUSINESS PARK | High-quality, low-density environment. Aimed at firms requiring prestige image or high caliber workforce. Mixtures of manufacturing office and sales function. | Aztec West, Bristol Silic, Rungis, Near Paris |
| UPGRADED INDUSTRIAL | Upgraded marketing image to standard industrial estate. Some building types with "high-tech" facades. | Aztec West, Bristol Silic, Rungis, Near Parks Stacey Woods, Milton Keynes Sutton Industrial Park, Reading |

As part of the early research program for Stockley Park, a 1.5-million-sq-ft business park near Heathrow Airport in London, DEGW undertook detailed case studies of how 26 firms used their current space, its shortcomings, and their future demands. The predominant lesson was the need for adaptability of the range of functions undertaken and the way the building was serviced and subdivided. Two examples reflect the need for flexibility: The first, a young rapidly growing firm that makes high-value, specialized, electronic mixing equipment for the television and film industries, moved into a laboratory building that they fitted out to a high office standard and now use for a broad spectrum of functions, with small pockets of subleased space to provide space for expansion. The second, an international microcomputer firm, took a standard institutional warehouse building, which it adapted by adding mezzanine space for office use. In

Typical manufacturing premises

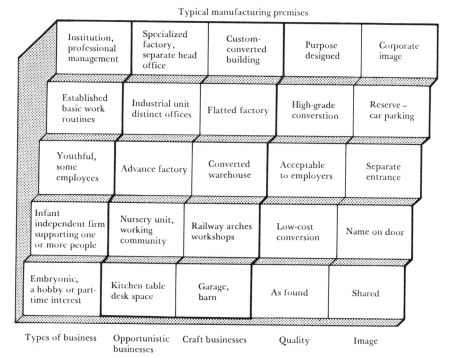

| Institution, professional management | Specialized factory, separate head office | Custom-converted building | Purpose designed | Corporate image |
|---|---|---|---|---|
| Established basic work routines | Industrial unit distinct offices | Flatted factory | High-grade converstion | Reserve – car parking |
| Youthful, some employees | Advance factory | Converted warehouse | Acceptable to employers | Separate entrance |
| Infant independent firm supporting one or more people | Nursery unit, working community | Railway arches workshops | Low-cost conversion | Name on door |
| Embryonic, a hobby or part-time interest | Kitchen table desk space | Garage, barn | As found | Shared |
| Types of business | Opportunistic businesses | Craft businesses | Quality | Image |

**Figure 3-9.** A firm's accommodation requirements vary according to the steps in its life cycle.

general there is pressure today for additional training and office space, and there is need for mechanized ventilation in offices, a better quality image, and more parking for staff and dealers.

## GENERIC BUILDING FORMS

We are all familiar with the accepted norms for industrial and office space. Offices, by institutional convention, are 40 to 46 ft (12.0 to 14.0 m) deep with floor-to-floor heights of approximately 10 ft (3.0 m) to allow for a suspended ceiling and servicing. Manufacturing and storage is deep single-storied space, with a portal frame to allow clear spans, and a dimension of 18 to 20 ft (5.5 to 6.0 m) to the underside of eaves. When the two types are combined, the components are separately identified in an architecture that was best reflected on the Great West Road in the 1930s. The work of architects such as Nicholas Grimshaw in the 1970s began to merge these differentiations by constructing a double-height volume within which the office space was built as mezzanines. Arup Associates, U.K. headquarters for Digital at Worton Grange, Reading (Figure 3-10), at-

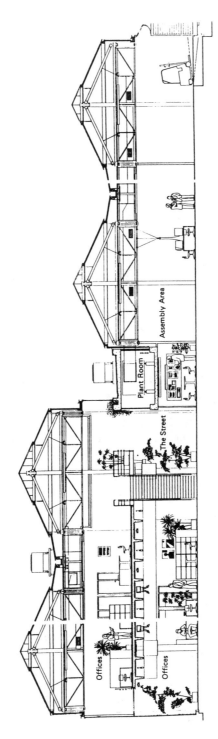

**Figure 3-10.** Mixed use high-technology environment: Digital Equipment Corporation United Kingdom, Warton Grange Reading. Architects: Arup Associates.

tempts to merge the demands of office, light industrial, and storage uses under one roof, with a concourse as the focal point, giving access to the various functions.

As a result of the demands for a flexible building shell within which differing functions can be effectively accommodated, two distinct shell types have emerged.

1. Double-height volume, deep plan single-storied building (18 to 20 ft height). The shell is attractive to larger firms (say 5,000+ sq ft or 465 sq m) requiring
   - space for high-technology process (height of servicing);
   - heavy machinery (requiring overhead gantries or forklift truck);
   - extra storage capacity (mezzanine);
   - proportion of office space that could be on an upper level;
   - expansion space within the basic building shell.

2. Single-height volume on one, two, or three floors, with shallow or medium-depth space on the upper floors to allow for natural ventilation and light. This type of shell is attractive to firms
   - requiring small tenancies, or ease of subdivision to units less than 5,000 sq ft;
   - undertaking bench-oriented high-value-added activities;
   - manufacturing or assembling small components or products;
   - using low-technology processes, with basic servicing requirements;
   - requiring maximum flexibility to change the relationship and proportion of different functions.

Our research for Stockley Park clearly identifies that the knowledge industries placed great emphasis on the need for a building that would enhance rather than restrict the operation of their business. They were concerned to achieve the following:

- Large uninterrupted floor areas, to allow for a mixture of functions that could grow and change with time.

- Medium-depth floors of from 56 to 60-ft (17.0 to 18.3 m) deep allowing for a variety of layouts, and a mixture of offices, group spaces, larger production areas, and a central well-serviced internal zone for equipment and support activities.

- Two- or three-storied buildings that may be zoned for security, allow for the flexibility of subleasing, and reduce internal travel distances to facilitate interpersonal and interdepartmental communications.

- Generous floor-to-floor heights (13 to 16 ft, or 4.0 to 4.9 m) that provide capacity for installing service ducts and cableways and allow sufficient headroom for specialized equipment.

- Capacity to install machines. Air handling for heat extraction required for machines rather than people. As information technology proliferates the demand will increase for finely zoned systems that can provide capacity for localized "heat spots" and generous and easily accessible zones for cabling.

- Special services for kitchens, cafeterias, training, and demonstration areas.

- Flexibility of tenure, through short leases and planning uses that recognized the need for the freedom to grow and change.

- Speed and flexibility to move in, by the developer providing a basic shell and fitting-out allowance for air-conditioning, ceiling, and lighting and floor finishes that can be rapidly installed to match the tenants' requirements.

- Customized buildings, where in association with the developer's team the predesigned shell could be adjusted and fitted out to meet specific requirements.

## CROSS-CULTURAL COMPARISONS

With the fluid planning laws in America, the precise distinction between industrial, office, and warehouse uses has been blurred, and the mixed-function building is acceptable. Some of the hurdles that we are seeking to change in Europe are not barriers in the United States. I hope, however, that describing European experience may help to sharpen American practice.

PART TWO

# PLANNING DEVELOPMENTS FOR HIGH TECHNOLOGY

CHAPTER FOUR

# LOCAL ECONOMIC DEVELOPMENT AND RESEARCH PARKS AND SETTINGS FOR HIGH-TECHNOLOGY ACTIVITIES: AN OVERVIEW

*CHARLES W. MINSHALL*

**O**ne of the most important characteristics of forward-looking development programs for a wide range of high-technology industries and services is the trend toward high-quality settings and facilities. In many parts of the United States, the development of science and high-technology parks by private developers, government agencies, and universities is designed to attract, develop, and retain desirable economic activities in order to enhance the overall attractiveness of the area.

This has become an important topic and has generated a competitive atmosphere between cities across the United States. As a result, criteria applied to both sites and structures are changing. Despite this clear trend toward high-quality settings for high-technology activities, confusion exists as to what constitutes a true high-technology park as opposed to a mixed-use office park or an industrial park. The following discussion will examine major trends in detail, including locational criteria, university affiliation, covenants, organization, and management.

Factors will be examined in each category that may contribute to either the success or the failures of settings intended for high-technology activities. These have emerged from the extensive studies carried out at Battelle Memorial Institute for such settings in various locations in the United States.

## HISTORICAL BACKGROUND

### Growth and High-Technology Activities

Today, high-technology activities represent an important source of new employment opportunities, a critical reason for developing appropriate facilities. Although the precise numbers differ, according to the definition applied, eight out of 10 high-technology activities have been identified as major growth sectors for the remainder of the 1980s and 1990s. When one examines the total universe of growth industries and services, a majority of new jobs created across all economic sectors involve a high-technology activity. Further, because most of the sites and structures appropriate for these activities are equally well suited to the rapidly growing service and office-related sectors, the development of suitable settings for both high-technology and administrative functions should be viewed as a mandatory component of any forward-looking growth program. Table 4-1 presents the results of a survey of activities expected to grow most rapidly in the 1980s and 1990s.

### High-Technology Parks

The first science or high-technology park in the United States was developed in Menlo Park, California, in 1948. Further development gained momentum with at least an additional 50 projects in the 1960s and another 25 to 30 in the early 1970s. In 1971, a survey by *Industrial Research* magazine identified 81 parks. Among them, 27 were restricted research parks and 54 were classified as industrial parks with an emphasis on scientific activities. Formal organization varied: 19 were affiliated with universities, 25 were nonprofit developments organized by community groups such as the chamber of commerce, and 54 were for-profit corporations. Location was concentrated in California, followed by Massachusetts, Maryland, Texas, and Colorado. The average size of the early developments was approximately 650 a. A number were considerably larger: notably, Sterling Forest in New York, 20,000 a; Research Triangle Park in North Carolina, 5,000 a; Huntsville Research Park in Alabama, more than 2,000 a; and University Research Park in Charlotte, North Carolina, almost 1,400 a.

Development continued through the 1970s, with a rapid increase in the early 1980s. By 1987 it was possible to identify well over 150 high-quality

TABLE 4-1
**Major growth activities of the 1980s and 1990s**

| INDUSTRIES | |
|---|---|
| Electronic components | Drugs |
| Aircraft and parts | Toilet preparations |
| Computers | Glass products |
| Fertilizers | Primary aluminum |
| Agricultural chemicals | Nonelectric heating equipment |
| Plastic materials | General industry machinery |
| Cement, clay products | Construction machinery |
| Printing | Mining equipment |
| Industrial controls | Industrial trucks |
| Radio, TV equipment | Special industrial machinery |
| Office machines | Electric motors, generators |
| Ordnance | Lighting fixtures |
| Pulp and paper | Scientific instruments |
| Paper containers | Medical, surgical instruments |
| Coal mining | Photo, optical equipment |
| Industrial chemicals | |

| SERVICES |
|---|
| Telecommunications |
| Radio, TV broadcasting |
| Air transport |
| Other transport services |
| Advertising |
| Motor freight |
| Water transport |
| Water, sewer services (infrastructure) |
| Business services (computers) |
| Medical services |

*Source:* Battelle National Input/Output Table, 1980–1990, 1990–2000 projections.

science- and technology-oriented developments, and probably an equal number of exceptionally well-planned projects ideally suited for a controlled mix of light manufacturing, business services, and office activities.

The rapid expansion of technology-oriented high-quality sites reflects the locational needs of many growth industries as well as the economic advantages to the communities that develop them. The development of sites and structures for high-technology activities must no longer be considered unique; in many respects, they represent the basic standard established by many communities in the United States.

## THE PROBLEM OF DEFINITIONS

Because no single definition can be applied to these evolving, new concepts, it is difficult to determine precisely how many exist and to assess their relative success. What one developer describes as a high-technology park may also be described as a research park, science-oriented park, industrial research park, industrial park, science center, technology center, or even an office park. The issue is further confused when a project, initially organized as a science park, begins to accept light manufacturing, offices, business services, and even certain types of distribution activities in order to increase the number of tenants.

A number of surveys have been conducted that help to determine the definitive factors; two general descriptors of *high-technology parks* have evolved:

1. That of a tract of land that has been developed with the necessary utilities, roads, and other site requirement on which buildings are constructed primarily for research and development activity to include prototype fabrication and some light manufacturing.

2. That of land devoted to research and development activities, with or without light manufacturing, corporate headquarters, warehouses, or office buildings, but with goals to foster interaction with a university. In fact, true high-technology parks may be considered a marriage between higher education and industry.

A brief overview of the differences between types of parks can be seen in Table 4-2. I will attempt to highlight significant factors in the following discussion.

## Science Parks

As the profile in Table 4-2 suggests, the major activities associated with this type of park include research, engineering, prototype development and certain types of office and administrative activities. In virtually every case, there are no plans to accept light manufacturing, distribution, or certain types of business services. Science parks are clearly oriented toward research, and their potential work force is dominated by professional and technical employees. In fact, several of the better-known science parks base their entrance requirements on the occupational structure of the work force. For example, in computer, electronic, instrument, drug, and pharmaceutical industries, 15 to 30% of the employees can be categorized as either professional or technical. A frequent requirement imposed on companies wishing to locate in a science park is that a minimum of 8 to 12% of the work force be professional—that is, in the scientific or technical category.

It is difficult, if not impossible, to identify successful "true" science parks. In most instances, restrictive rules or covenants have decreased the demand for the pure science park, contributing to many of the failures of

**TABLE 4-2**
**Scientific/industrial park continuum**

| | SCIENCE PARKS | TECHNOLOGY PARKS | OFFICE/ MIXED USE | "AAA" INDUSTRIAL | INDUSTRIAL |
|---|---|---|---|---|---|
| Occupational structure: emphasis on | | | | | |
| Professional: technical and scientific jobs | Yes | Yes | Some[a] | No | No |
| White collar: administrative, clerical | Some | Yes | Yes | No | No |
| High skills assembly: craftsmen | No | Some | Yes | Yes | Yes |
| Other assembly: operatives | No | No | Some | Yes | Yes |
| Services | No | Some | Yes | Yes | No |
| Distribution, transportation | No | No | No | Yes | Yes |
| Major activities | | | | | |
| R&D—R&E | Yes | Yes | Some | No | No |
| R&E, High Technology Production | Some | Yes | Some | No | No |
| Production—End Items | No | Some | Yes | Yes | No |
| Production—Components | No | Some | No | Yes | Yes |
| Distribution | No | No | Some | Yes | Yes |
| Office-Administration | Some | Yes | Yes | No | No |
| Industry orientation | | | | | |
| Technical orientation: instruments, electronics, drugs, machinery, transportation equipment | No | Yes | Yes | Yes | No |
| Diversified: chemicals, printing, plastics, fabricated metals | No | Some | Yes | Yes | Some |
| Production oriented: food, textiles, apparel, wood, paper, furniture, petroleum, stone/ glass, primary metals | No | No | Some | Yes | Yes |
| University Tie | Yes | Perhaps | No | No | No |

[a] Some means that the activity may be accepted depending on the nature of the facility, processes, and applicable covenants.

both university and private sector science parks. It has been estimated that fewer than 300 independent research and development centers, employing more than 100 people each, exist in the United States. Because more companies are linking research, pilot, and production activities, the demand for the "pure" science park is declining.

## Technology Parks

The technology park may be characterized as a quality development suited to a wide range of activities, from research and development to high-technology and light manufacturing activities to office and administrative functions and services. Here, scientists, engineers, and technicians may work in large numbers along with white-collar professionals, clerical, and highly skilled production personnel and computer-oriented operatives.

A relatively large number of successful projects, with a diversified range of activities, may be defined as high-technology parks: among them, two oft-cited success stories, the Stanford Research Park in California and the Research Triangle Park in North Carolina. Characteristically, their design and development are controlled by stringent covenants. An essential ingredient often associated with high-technology parks is the direct support by a local university. Without a strong relationship between higher education and industry to nurture research and development activities, the typical park has demonstrated a propensity to fail or shift to a mixed-use status.

## Office Parks

The major differences between technology parks and the more general office/mixed-use developments appear to be qualitative and philosophical. The office park tends to provide not only for research and development but also for a wide range of office, light manufacturing, and business support services as a result of less rigorous covenants.

## Industrial Parks

Two basic categories of industrial parks—standard and AAA—are clearly oriented toward traditional production, service, and distribution and are not well suited to a wide range of high-technology activities.

## Summary

As can be seen from the above definitions, it is often difficult to differentiate between the various categories of development. In many cases, the major difference may be the quality of the project and/or the covenants that govern them and not the label attached to the development. High-quality industrial parks will remain attractive locations for many types of technology-oriented production activities and some services. Those with less rigorous covenants and undesirable activities may preclude the participation of potential high-technology tenants. In summary, there are no clear dividing lines between categories.

Increasingly, high-technology activities are being attracted to high-quality settings especially suited to their particular needs: those that include a work environment conducive to the attraction and retention of high-caliber employees along with both basic and support activities. The following section will discuss in some detail the criteria conducive to such high-quality settings. Lessons learned from recent history will pinpoint the dos and don'ts that appear to be causal factors in determining successes or failures.

## CRITERIA FOR SUCCESS OR FAILURE

Successful parks do not appear overnight, but like good wines develop and mature over many years! The best-known parks are relatively old, long-term projects. For example, expansion in scientific and technical activities along Route 128 in Massachusetts occurred in response to needs generated by World War II. The highly regarded Research Triangle Park in North Carolina traces its origins back to the 1940s; Stanford Research Park in California opened to tenants in 1951. However, developers who think success lies in attempting to replicate these well-known projects instantly are likely to be greatly disappointed. Current economic, political, social, and international conditions are markedly different from those of two or three decades ago.

The historical growth of successful parks is related to land absorption rates. According to an analysis of 27 high-quality research parks (most allowing mixed uses), approximately two new tenants are attracted per year. A boom period often occurs after the first one or two tenants have been attracted (anchor tenants are critical), and four or five new firms may be attracted. Similarly, some projects demonstrate considerable momentum and may attract three or four new tenants for several years. Still, in most cases two or three tenants per year means success—and these involve a total of 11 to 13 in most projects.

It is possible to learn from past successes and failures, as discussed below: the most successful science and technology parks share such characteristics as a parklike setting, judicious application of covenants, and direct participation and support by a local university and business and support centers. Some of the lessons learned pertain more specifically to university-related science and high-technology parks. In a survey of such parks built between 1960 and 1980, Battelle found that over three out of four were not successful. Some projects were abandoned, others were utilized for normal university expansion, and a few were sold off and devoted to more traditional office and production uses. What has become very clear in the past 3 or 4 yr is that universities have learned a lot about the causes of failure and the criteria for success, and the projects that exist now, or are evolving, are much different from those of the past. Homework has been very important here, and at least one university—before even committing to a science park project—sent senior staff to visit over 40 projects across the

United States in order to clarify the most appropriate strategy to be taken. Where specifically concerned with university-related parks, the following discussion of major characteristics associated with successful parks utilizes a survey completed by one major university in the Mid west.

## Decision Makers and Team Organization

Today, many of the decision makers in high-technology activities are the scientists and engineers themselves. It is not unusual for senior technical managers to be asked to select a location and arrange for the structures to house a production or research facility. The qualitative aspects of the potential setting are important factors in the decision process, from the general layout, landscaping, density, setbacks, contiguous activities, and existing roads to construction materials. Existing buildings have emerged as a major element in the process, both in attracting and retaining tenants; they save time and money. High-quality facilities, those that offer flexibility, special utilities, relevance for the specific operation, and reasonable costs, are prime prospects.

Several recent studies concerning the locational and structural concerns of serious prospects indicate the same pattern: fewer than two out of 10 are interested in purchasing undeveloped land; fewer than three out of 10 are interested in even a high-quality site to build on. At least 70% want to be able to choose among several alternative locations in a particular community appropriate to scientific and technical activities. Accordingly, available *starter,* or *incubator,* spaces and their related services are perhaps more important than other aspects of development. The average new industry wants 30,000 to 40,000 sq ft (2,787 to 3,716 sq m) of space; the average office requires 15,000 to 25,000 sq ft (1,393 to 2,322 sq m). Given the competition for increased economic activity, cities or developments that do not have space or facilities available are virtually eliminated from further consideration.

Where parks are linked to universities, a clear trend exists to hire professional developers and managers to market and control their projects on a day-to-day basis. Major problems have arisen when this mandate was given to academic or administrative staff on a part-, or even a full-time basis— largely because of their lack of experience. While the degree of independence of parks varies from university to university, it is clear that professionals are being entrusted more than ever with saving presidents from the hassles of operational problems, providing not only professional guidance but also insulation between the operations of the project and the university staff, encouraging aggressive marketing, and encouraging development through mechanisms such as profit sharing and incentives (not available usually to state employees). In other words, the professionals are beginning to take over.

The most successful parks have been characterized by very tight professional management, clear organization, and intent that spells out allowable uses of individual properties, guidelines for new activities, and specific planning. (In most cases, the owner or organizer had a clear role in establishing the park and covenants, ensuring services and maintenance, and monitoring day-to-day activities.) Additionally, many have a clearly identifiable "authority" or "board," a full-time director and staff, a development-review committee, and an association of tenants.

The organization of successful parks reflects, significantly, a team effort. These parks are characterized by extensive professional marketing (advertising, direct mail, public relations, and electronic media) usually coordinated by the full-time executive director. In a majority of them, efforts are made to coordinate the marketing of the technology park with the local chamber of commerce, the state department of development, and private sector developers representing utilities, realtors, bankers, etc. In other words, the key to success involves a tight and well-organized management, a highly professional and full-time executive director and staff, and a marketing program that extends across the community.

## University Ties

Perhaps the most important factor that differentiates the true high-technology park from the high-quality office or industrial park is the support of or link to a university. This topic is discussed at length in Ira Fink's Chapter 5 in this volume.

Many benefits accrue to the park from the support of a university with strong graduate programs in math, engineering, the sciences, business, and management. Directly or indirectly, the programs contribute to the success of the high-technology park. Benefits include faculty expertise, the availability of graduate students for selected research jobs, the opportunity for park-related researchers to become adjunct faculty members, personnel screening and support access to professional seminars, symposia, and conferences (to keep technical/scientific staff up to date), professional growth opportunities through advanced degrees or continuing education (absolutely essential given the rapidity of technical change), contract research support, extensive library resources (including the various computerized search and keyword systems), shared facilities and equipment (computers and laboratory devices), as well as the social, cultural, intellectual, and recreational benefits inherent to the university setting.

Benefits also accrue to the university. The increased opportunity for employment in a "real world" setting for faculty, graduate students, or spouses is an added bonus, while the opportunity to employ selected employees as professors or instructors contributes to the overall quality of education.

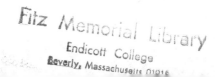

The role of the university should not be considered "icing on the cake." Among university-related parks across the United States, over 80% of all projects demonstrate a strong concern for university/industry ties—supported by definite programs for students and others. Only one or two major projects downplay this thrust, and while they have been successful in moving land, they are really office parks with a university name. While virtually no university would impose constraints related to minimum levels of contract research, hiring, or other ties on prospects, the desire to continue meaningful ties is important—and sets these projects off from other real estate developments. Applied with reason, encouraging ties seems to help more than hinder.

However, a number of world-class universities across the United States have not led to the development of high-technology parks or even a significant number of spin-off activities. This fact has increasingly enhanced the reputation of those that have supported such projects while offering acceptable science, engineering, and technical programs.

## Economic Climate

Developers of successful technology and science parks recognized from the start that an aesthetically pleasing site would assist in attracting tenants, but realized that it would be inadequate in and of itself. In evaluating a site, the potential tenant considers additional geographic, environmental, and economic factors, including:

- Positive overall business climate in terms of incentives, costs, labor/management relations, etc.

- Areas characterized by reasonable and acceptable cost, not only those related to operations (such as taxes and labor) but also those associated with individual households such as salaries, personal and real estate taxes, the cost of living, and other concerns such as good sites available on affordable financial terms.

- Flexibility to expand at reasonable cost (in and near the park)

- Absence of mixed land uses in or contiguous to the park

- Proximity to high-quality air transportation

- High-quality residential areas within a reasonable commuting distance with abundant housing choices and reasonable costs

- Areas characterized by very good elementary and secondary schools (both public and parochial)

- A wide range of cultural and recreational activities available in the community

Also, particularly for university-related parks, an important decision concerns the lease or the sale of land. Although a significant amount of diversity exists here, more than 50% of the projects are concerned primarily with leasing land and facilities. Approximately 15 to 20% are concerned only with sales, and approximately 33% will consider either. Overall, leasing is preferred and is seen as one way in which the university may continue to influence the long-term nature of the project.

## Spin-off Potentials

Note that many high-technology parks often grow "from within" through the establishment of spin-off companies that are either research and consulting firms or production-type organizations that capitalize on new products or processes. The spin-off companies are formed by individuals who draw heavily on scientific and technical knowledge used in their previous employment with parent companies, universities, government research agencies, or nonprofit laboratories. Often the individuals responsible for the spin-off firm feel boxed in by company or university policies, draw upon ideas generated on their own time, buy or are given rights to inventions that their employers did not want to pursue, or are simply entrepreneurs who wanted the experience of operating their own company. What is important is the need to recognize that the spin-off firm represents great growth potential for the typical high-technology park. And conversely, the park environment is much more likely than other environments to generate such spin-off activities. Further, it has often been noted that a "chain reaction" may be associated with spin-offs. It is highly desirable to have a wide range of incentives and supporting services available to facilitate this type of activity, such as:

- Specific financial programs including seed money, venture capital, programs of the small business investment corporations, and business development corporations

- Incubator space at low cost either in older buildings or in shared space in shell-type buildings with movable partitions and a wide range of shared equipment such as computers, laboratories, and libraries (see Davidson's Chapter 6 in this volume)

- Research contracts and guaranteed sources of income to assist the company in its formative stages

- Specific incentive and assistance programs provided by local chambers of commerce and development organizations not only in the technical area but also in areas such as legal, patent, and business planning

Thus, in addition to very intensive efforts designed to attract new high-technology industries from other states and regions and the even more important programs to be implemented to facilitate the retention or attraction of industries from within the metropolitan area or state, the forward-looking industrial park developer will also consider the development of specific programs to foster spin-off companies. A majority of the types of incentives and facilities required for the spin-off inducement program involve incubator and starter space.

## Land Use, Activities, and Support Services

### LAND USE

Land use patterns will be controlled at the two levels of local zoning ordinances and of covenants regulating the activities permissible in a particular park (see below). Very few projects allow distribution or warehousing, and virtually none are interested in heavy manufacturing. A majority allow light and technology-related production, research, development, prototype and pilot operations, and office activities. The emphasis is on controlling the activities of tenants through covenants—not on the wholesale exclusion of activities by type.

### TENANT SERVICES

One of the most important trends is that the projects are providing many more services for tenants, including conference centers and meeting space, finished lease space in multitenant buildings, incubator space, or direct ties to university services where relevant.

### PERSONAL SERVICES

In addition there is a definite trend for projects to encourage the attraction of personal services to their sites or to their peripheries. These include motels, fast food and other eating establishments, banks, bookstores, exercise facilities, and general merchandise stores. For university-related parks, a second perspective is very important. This is the fact that a number of professional and support services are being made available, as discussed above. Also, managers and employees may be allowed certain university privileges, such as access to swimming pools and tennis courts.

### BUSINESS SUPPORT

Just as there has been a progressive shift from unimproved sites to AAA industrial parks, to technology parks in parklike settings, a new type of development is emerging. The business support center is often located in the city center, often in a single office-type building, providing a mix of space and support services, research, and even light high-technology production. In other words, any type of activity that does not cause noise, dirt, traffic, or other detrimental factors may be suitable. Access to a business support center that in addition to high-quality space offers complete secretarial, computer, and copying services has become an important criterion in location. Others also offer some tenant services, as described above.

## INCUBATOR

The business support center is especially well suited to provide *incubator space* for new business and emerging high-technology activities (see Davidson's Chapter 6). Now, over half of the university-related projects provide incubator support, but this share is increasing rapidly. Several universities filled their incubators within weeks of opening, and have developed long waiting lists. Most are in the range of 35,000 to 80,000 sq ft (2,351 to 7,432 sq m) in size, and offer a complete range of services, such as dry labs, computer, secretarial, basic business planning, staff, and other university support, some financial support, security, furniture, and certain types of equipment.

## Aesthetics and the Physical Environment

Many of the criteria associated with high-technology parks revolve around the concept of a real "park." Campus settings, aesthetics, and green space are all descriptors that have been applied to the most successful projects; they are attractive, nice places to work. A number of highly attractive buildings are developed in close proximity to each other. Desirable, competitive developments have vast areas of green space, streams, trees, off-street parking, and buried utilities. Buildings are characterized by attractive exterior finishes, elaborate heating and air-conditioning systems, office-quality lighting, wall and floor coverings, high-quality, flexible interiors, telephone/communications hookups, computer text processing cable runs, light-duty laboratory space, and common areas. Prospective tenants are more quality conscious than ever. High-technology production, service, and research and development activities are increasingly attracted to the campuslike settings described earlier.

For university-related parks specifically, the survey mentioned earlier also highlighted some of the quantitative factors that contribute to the campuslike settings:

- Building to Land Ratios
  With the emphasis on green space and campus settings, these successful projects demonstrate very low densities. Overall, between 25 and 30% of a site may be covered by buildings, and less than 50 to 60% will be utilized for buildings and parking lots.

- Lot Size
  In a majority of cases, minimum lot size is specified but tends to be relatively small: 3.5 to 4 a on average. Approximately one-third of all developments specify maximum size, usually 20 to 25 a. This latter constraint is flexible, with much larger sites being made available to "blue chip" companies with large employment.

- Number and Size of Buildings
  Few projects are concerned with the number of buildings per site, but through the use of coverage and other covenants ensure that acceptable densities prevail. Similarly, building size is highly variable. Some projects have buildings with as few as 5,000 sq ft (464 sq m), and others exceed 1,000,000 sq ft (92,900 sq m). Overall, the average building is approximately 65,000 sq ft (6,038 sq m) in size. Further, while most are single-story structures, there is a definite trend to have two- or three-story structures with very fine architectural design. The vertical development is especially important for those projects located in built-up areas with limited site size. Here, building up is essential if one is serious about maintaining the green space in the project.

- Number of Tenants per Building
  Surveys by several universities and technology park managers suggest that over 50% of tenants require their own buildings, at least 25% desire single-user space, and only around 25% have no preference in this area. This trend may be at variance with the pattern noted above to move toward multistory structures. Also, a major exception to the pattern just noted is found in regard to incubators where multiple tenants are expected and to some extent encouraged.

## The Covenants—Keys to Success

The application of covenants distinguishes the science or technology park from traditional industry and service sector sites. In many cases, the success of a technology park depends upon covenants applied to land use, landscaping, aesthetics, permitted activities, structures, traffic, pollution, and utilities. Characteristically, the most successful parks have established and enforced a wide range of rules, while not limiting arbitrarily the activities they will accept.

Typically, the covenants include a description of the project, definitions, general provisions, variances, permitted land uses, performance standards, architectural and engineering factors, and review procedures. They may include guidelines regarding access, space allocation, site coverage, setbacks, expansion, utilities, landscaping, outdoor storage, parking, loading, materials, colors, signs, emissions, noise, employment patterns, and mandated services.

## ARCHITECTURAL AND ENGINEERING ASPECTS

A more detailed look at the architectural and engineering factors employed by successful projects reveals a pattern of strict adherence to the rules regarding the use of materials, colors, signs, utilities, landscaping, outdoor storage, etc. For example:

- Materials
  Structures must be finished on all sides, preferably in masonry, brick, stucco, and selected forms of aggregate. Metal-clad buildings, metal roofs, and wood frames are undesirable.

- Color
  Earth tones should predominate to blend with the environment. Those that clash or are simply too bright—i.e., blues, reds and yellows—should be avoided.

- Signs
  These should be mounted and be as small as possible for identification only. No advertisements or billboards are allowed.

- Utilities
  All connections and lines are underground. Related equipment, such as transformers, are completely screened.

- Landscaping
  This is extensive and based on professional plans. Successful projects go to great lengths to facilitate an attractive physical environment. Typically, berms and green belts are incorporated. Trees are not removed unnecessarily, and in many instances groves of trees are planted to block unwanted views, structures, or activities. Wherever possible, efforts have been made to develop or preserve ponds, creeks, rock outcroppings, and even open meadow spaces; walking and jogging paths make them accessible to employees.

- Outdoor storage
  This is strictly forbidden, as is the use of temporary buildings. Unsightly but necessary structures or equipment, such as dumpsters, are screened from view permanently.

## SPACE ALLOCATION

In contrast to the large, successful high-technology park, the average park is of a smaller scale. Resources are concentrated on making the best use of the site available, often utilizing the campuslike setting where buildings are developed in close proximity to one another. A closer look reveals the following patterns:

- Minimum size
  The requirements in most successful high-technology parks are in the

neighborhood of 3 to 4 a. Some of the newer developments require from 5 to 6 a. While the larger sites provide a spacious feeling, developers should be aware that the higher costs associated with them could be counterproductive.

- Coverage
  This is strictly controlled. In most cases, it should not exceed 50%—that is, no less than 50% of any particular site may be utilized, including all building and paved (access, parking) areas.

- Setbacks
  These are rigorously enforced. They usually require a minimum of 50 ft (15.2 m) from the front of the property, 25 to 35 ft (7.6 to 10.7 m) from the back, and 30 to 40 ft (9.1 to 12.2 m) from the sides, with a minimum of 100 ft (30.4 m) between buildings. In most developments there are guidelines regarding site configuration.

- Space allocation
  In the more successful high-technology parks is typically as follows:
  - Streets, roads—10%
  - Institutional—5%
  - Common/open areas—10%
  - Uses: Mixed: office/industry—75%; Segregated: Office 25%, industry 50%; Office 30%, industry 35%, commercial 10%; Office 20%, industry 15%, commercial 10%, residential 30%

- Heights
  These are usually restricted to a maximum of 45 to 50 ft (13.7 to 15.2 m). In most cases, no overt pressure is applied to construct single-story buildings, and multiple-story buildings are acceptable. The impact of excessive height may be reduced through the use of extensive landscaping and berming.

- Parking and Turnarounds
  The rules on parking are clear. Absolutely no parking on streets is allowed. All parking areas must be paved. Paved areas do not count in calculating required setbacks or open spaces. Paved parking areas require islands and barriers and in many cases must not be visible from nearby streets. Loading and unloading must take place on site. The requirements for allocating parking spaces are quite rigorous, for example:
  - Traffic aisles should be a minimum of 18 to 22 ft (5.5 to 6.7 m).
  - A landscaped island may be required for every five to 10 cars.
  - One parking space would be required for each 200 to 300 gross sq ft (18.6 to 27.9 sq m) of built space.
  - For laboratory activities, a minimum of one space would be required for every three employees.

- Permitted Land Uses
  In almost every case, considerable flexibility exists in regard to land use and the activities allowed. Activities that tend to be excluded involve warehousing and distribution because of the excessive traffic they generate, along with heavy manufacturing and those industries related to the processing of raw materials, such as wood products.

- Occupational Structure
  As mentioned earlier, covenants may also regulate acceptable occupational structure. An examination of employment patterns in high-technology parks reveals that activities related to research, development, pilot prototype plants, and (certain types of administration related to research) management are important and may account for between 25 to 35% of all jobs. Between 35 and 55% of all jobs are associated with light manufacturing and high-technology production. The remaining employment categories tend to be quite diverse: the most common being office and administrative, followed by personal and business services.

- Performance Standards
  To ensure the high quality of the technology park, rigorous performance standards are applied to all tenants. They can include the following: *noise,* must be minimal, measured in specified decibels; *smoke,* essentially none allowed, measured on Ringleman charts; *particulates,* essentially none, measured with a Millipore Particulate Counter; *toxic gases,* none; *fumes,* none; *vapors,* none; *vibration,* minimal, measured in cycles and frequency; *glare,* none, or must be screened; *lighting,* controlled, should be directed internally; *effluent,* must meet local, state, and federal standards; *discharge* must be carefully controlled; *waste,* no external storage, formal removal plants are required; *radiation* is virtually not allowed or must be screened; and *fire and safety,* no bulk storage of any dangerous material. Note that the burden of compliance with the covenants is upon the tenant.

## CONCLUSION

Table 4-3 presents a summary of major high-technology park characteristics. Factors that have most contributed to the success of high-technology parks follow:

- They are characteristically located on large suburban tracts (at least 200 a, but preferably between 350 and 750 a), with specified low densities, setbacks, and requirements related to off-street parking—all space-consuming needs.

TABLE 4-3
**Summary of major high-technology park characteristics (based on an overview of 32 projects across the United States)**

| CHARACTERISTIC | SCIENCE PARKS | TECHNOLOGY, MIXED-USE | LIGHT INDUSTRIAL |
|---|---|---|---|
| Average size | 125 a | 477 a[a] | 510 a[b] |
| Percent available | 95 | 80 | 60 |
| Average number per city | Less than 1 | 3–5 | 5–8 |
| Activities accepted Main | | 98% R & D 94% Offices 87% Light prod. 14% Distribution 47% Services 22% Manufacturing | |
| Supporting services | | 64% Fast foods 47% Motels 38% Other retail 60% Banks & automated tellers 40% Conference centers 37% Recreational | |
| Number of tenants | | 37 approximately in successful developments (need for good anchor tenants) | |
| Average age | | 12 yr (successful projects) 20 yr (most prominent) | |
| University affiliation | | Stressed in 38% of projects Mentioned in an additional 24% | |
| Start-up assistance | | 55%: Projects with starter space and business assistance programs 8%: With viable incubator for high-tech firms | |

[a] One third are over 500 a, one in five over 1,000 a.
[b] One in 10 are under 100 a, six of 10 are between 250 and 600 a, one in 20 exceeds 1,000 a.

- Development periods typically have been much longer than previously anticipated—in virtually every case a minimum of 20 to 25 yr.

- Many parks have well-developed ties to major universities with either recognized science and engineering programs or an interest in establishing industry-outreach programs.

- They have well-developed links to a wide range of high-technology activities in the region and to a lesser extent to traditional industries.

- They offer a number of unique incentives: financial inducements ("up-front" or seed money for site acquisition and improvements, attraction of appropriate tenants), starter space, existent buildings, and many types of support for scientific and high-technology functions, such as computer support.

- They are characterized by very flexible pricing structures, mixed-land uses, and environmental constraints that are strictly enforced by tight management.

- They have been aggressively and professionally marketed by economic development groups or private developers to attract and spin off high-technology activities. The park's management works very closely with traditional development groups such as the chamber of commerce. The park may have quasi-public ownership, with a board of directors that involves government, university, developers, and the private sector.

From the point of view of businesses that locate in a technology park, the following advantages have been realized, according to recent surveys.

- Availability of incubator space

- Availability of access to an established labor pool

- Availability of highly organized and dependable services (maintenance, snow removal, catering, etc.)

- Availability of established utilities and infrastructure

- Well-defined and identified opportunities for expansion

- An aesthetically pleasing environment, enforced by very powerful covenants, which tends to stimulate creativity and contributes to the happiness of workers and together with the park's location makes the recruitment and retention of personnel easier

- Provision of a wide range of information about the specific community by the park developers, who also served to expedite the relocation process by taking care of local coordination and cutting red tape

- Availability of a wide range of tenant and personal services

These extra services not only serve to please employees and to induce them to work for the high-technology park's firms but also contribute directly to making the development more appealing. Considerable competition exists nationally to attract and retain scientists and engineers, and to draw such professionals to firms within the high-technology park. It is helpful to ensure that their desired life-styles are satisfied to the maximum extent possible and that a number of intellectual and financial rewards are made available. Good schools and housing and intellectual, recreational, and cultural offerings are often important factors in locating. In fact, some scientists and engineers consider the precise nature of the job less important than the community where it is located and, to some extent, the setting of their employment.

Finally, the objectives will, to a large extent, influence the operation and success of a project. For example, in a quasi-public project, the objective of the undertaking may be to facilitate job creation. So public ownership of the development may be critical, and this may facilitate marketing the project by local chambers of commerce, utilities, developers, and others. Also, for a park to be successful, prices for land, facilities, and services should be low, and public participation may be needed to support the trade-off between profits from real estate and income generated by new jobs.

## BIBLIOGRAPHY

Finholt, Rick. *A Comparative Study of University Affiliated Research Parks:* Prepared for the President, Ohio State University, 1984.

"Business Takes Root in University Parks." *High Technology:* January, 1986.

"Little Silicon Valleys." *High Technology:* January, 1987.

"Locational Preference of Scientists' Households." *Industrial Research:* January 1983.

Minshall, Charles W. *An Overview of Trends in Science and High Technology Parks:* Battelle Occasional Paper No. 37, 1985.

"University Research a Plus For Business." *Plant Sites and Parks:* September/October, 1986.

*The Urban Land Institute, Development Review and Outlook, 1984-1985.* "Research Parks, an Emerging Phenomenon." pp. 89–99.

U.S. Bureau of Labor Statistics. "Current Employment and Projected Growth in High Technology Occupations and Industries," Preliminary Report for the Joint Economic Committee of the U.S. Congress, August 2, 1982.

U.S. Bureau of Labor Statistics. *Occupational Projections and Training Data.* U.S. Government Printing Office, Washington, D.C., 1982.

CHAPTER FIVE

# THE ROLE OF LAND AND FACILITIES IN FOSTERING LINKAGES BETWEEN UNIVERSITIES AND HIGH-TECHNOLOGY INDUSTRIES

*IRA FINK*

## OVERVIEW

The subject of the role of facilities in fostering high-tech linkages between industries and universities is quite broad. To narrow it somewhat this chapter will concentrate on two areas: first, an overview of university/industry research linkages, and second, examples of how these linkages are occurring in various university and university-affiliated facility types. These facility types will be broadly defined to include space used by either universities or industries, including jointly owned facilities, innovation or incubator centers, supercenters, and university-affiliated research parks.

This chapter has been reprinted with permission. It originally appeared in a publication by the Society for College and University Planning in 1985.

In looking at university/industry research facility linkages, there are three basic points to be made:

1. In comparison with funds universities spend themselves for research or facilities, little of the applied research or facility money of industries is spent with universities.

2. One vital aspect of the linkage, which in effect is a form of technology transfer, is student recruiting.

3. Significant facility types that allow this technology transfer to take place are now found in university-affiliated research parks, in innovation and incubator centers, and in jointly sponsored facilities, or what can be called middlemen, or supercenters.

Underlying these program linkages are multiple factors: The first relates to basic goals of most research activities between universities and industry: to create new knowledge, mainly through research; and to help industry develop new products through technology transfer.[1] The second involves a shift that is under way as universities increase their research capabilities, particularly in high-tech areas. Historically, universities have been centers of teaching, with basic research an integral part of instruction. It is only in their recent history that they have also become major centers of applied research. As a result, within universities there appear to be two groups—the gatekeepers of the traditional values and the entrepreneurs.[2]

The third is recognition that inventions that create great new industries do not happen very often. While inventions such as plastics in the fifties, computers in the sixties, semiconductors in the seventies, and biotechnology in the eighties are epochal events,[3] there is always the possibility that the next one is awaiting discovery. And finally, the program and facility linkages that bring the participants together are often complicated because the motivations of universities and of industry are so different and the differences are so basic.

# RESEARCH RELATIONSHIPS

## University/Industry Research Linkages

The research linkages between universities and industries are actually only two parts of a three-part relationship. The third partner is government, which in the past was the federal government, but today also includes state or local government.

Historically, the research relationship between universities and industries has not remained constant.[4] For example, during World War II, the war effort sparked close ties between the two. During the 1950s and 1960s the federal government played the major role in funding of research and facilities, and the bond between universities and industries was somewhat weakened. During the 1970s and continuing today, with reduction in federal government funding for research and support facilities, universities have increasingly reached out for private sector as well as state and local government support. These contacts have increased in recent years as the business community has become interested in ways university resources can be employed for the benefit of both parties. This includes efforts at regional economic development, as well as commercial product development.

Figure 5-1, taken from the study *Report on Fostering Emerging Technologies in Texas,* illustrates the three components in the high-tech system of university, industry and government.

The primary purpose of university research remains to support education and the free pursuit of knowledge. Requests for facilities are most often initiated by the university. University policies on faculty consulting and the entrepreneurship of the faculty have a direct bearing on the extent of facility and research programs between universities and industry.

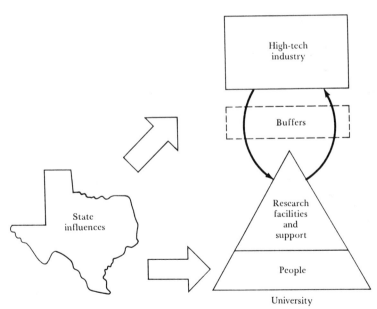

**Figure 5.1.** The high-tech system (source: "Report on Fostering Emerging Technologies in Texas," Advisory Committee to the Texas Senate Committee on Business, Technology and Education, December 1984).

## Technology Transfer

In broadest terms, one linkage between universities and industry, whether it is through facilities or otherwise, is technology transfer. As noted by economist Robert Premus, the most effective means of technology transfer from universities to high-tech companies is through student recruiting. As shown in Table 5-1, two thirds of the high-tech companies surveyed by the Joint Economic Committee of Congress found student recruiting to be the most important mechanism of transfer of technology. University publications, including books and articles, were considered important by almost one half of the companies. University services were important to one third. Corporate support of basic research was important to only one quarter of the companies as a mechanism for technology transfer.[5]

While student recruiting is an effective means of technology transfer, the institutions themselves must have viable research programs and facilities that are attractive to faculty and scientists, and in turn to students. The "bootstrap" effect of state actions is illustrated in Figure 5-2, which describes the circular effect of the relationship between funding, research, discovery, and linkages. Thus, one primary objective of industry in establishing university research or facility linkages is to obtain competent and adequate personnel. The actual need for specific research often lags behind as a primary industry objective.

## INDUSTRY

### Industry Locations

An important point to remember with regard to facility linkages is that most industries operate in a limited number of locations or with a limited number of facilities. It has been claimed that high-tech firms are "foot-

---

TABLE 5-1
**Technology transfer mechanisms for universities to high-tech companies**

| ALTERNATIVE TRANSFER MECHANISM | VERY IMPORTANT OR IMPORTANT (%) |
|---|---|
| Student recruiting | 67 |
| University publications | 46 |
| Faculty consulting | 42 |
| Government distribution of basic research results | 31 |
| Corporate support of basic research | 25 |

*Source:* Premus, note 5.

loose" in their location choice. However, as shown in Table 5-2, in a nationwide survey sponsored by the Joint Economic Committee of Congress, 40% of the high-tech companies had only one plant. About one third had five or more plants.[6] Stated another way, it is apparent that many high-technology companies are locally based, homegrown industries. Only when they are successful do they branch out to more than one location. During the start-up years they are likely to be located near the place they were founded, whether it be in the university setting, in someone's garage, or in inexpensive space that permits more of their limited funds to be used for research and development.

## Industry Research Expenditures

Because there is no published nationwide data on facility expenditures between universities and industries, research expenses will be used as a sur-

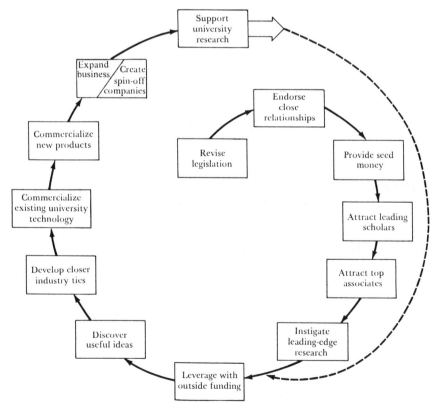

**Figure 5.2.** "Boot Strap" effect of state actions (source: "Report on Fostering Emerging Technologies in Texas," Advisory Committee to the Texas Senate Committee on Business, Technology and Education, December 1984).

**TABLE 5-2**
**Number of plants operated by high-tech companies**

| NUMBER | PERCENT |
|---|---|
| One | 40 |
| Two | 13 |
| Three | 9 |
| Four | 6 |
| Five or more | 32 |
| TOTAL | 100% |

*Source:* Premus, note 6, p. 22.

rogate to explain one point. In 1983, it was estimated that total national spending for research and development was $97 billion. As shown in Table 5-3, the share of this amount that represented industry investment in academic research was approximately one half of 1%, or about $425 million.[7] Thus industry investment in academic research was less than one tenth of what the federal government provided to colleges and universities and less than one quarter of what the colleges and universities were able to obtain themselves from state and local government funds.

Spelled out in another way, industry contribution to universities and colleges for research and development amounts to less than 1% of all money that industry spends for research and development itself. In other words, $99 out of every $100 that industry spends for research and development is spent within the industries. It would be safe to say the same holds true for facilities.

As one report indicates, among the 200 leading research universities in the United States, only 25 obtain more than 10% of their research funds from industry.[8] Among newly formed high-tech companies, it is unlikely that much, if any, of their highly prized capital would go to universities.

## RESEARCH LINKAGES

### Types of Research Linkages

As shown in Figure 5-3, high-technology development relationships occur on a continuum beginning with objectives and extending through funding before they occur at the university level that provides personnel, environment, and facilities, supported by outside funding agencies. From this background the program and research relationships occur.

In this regard, there are many categories of program linkages in support of technology transfer: each involves a different facility arrangement.[9] These include, for example, consulting, research grants, major contracts, affiliate programs, consortia, cooperative arrangements, exchanges of

**TABLE 5-3**
**Research and development (dollars in billions)**

| | | | PERFORMERS | | | | |
|---|---|---|---|---|---|---|---|
| SOURCES OF FUNDS | FEDERAL GOVERNMENT | INDUSTRY | UNIVERSITIES AND COLLEGES | ASSOCIATED FFRDC'S | OTHER NONPROFIT INSTITUTIONS | TOTAL | PERCENT DISTRIBUTION, BY FUND SOURCES |
| Federal government | $10.8 | $23.4 | $5.5 | $2.8 | $1.8 | $44.3 | 45.7% |
| Industry | — | 48.6 | 0.4 | — | 0.3 | 49.3 | 50.9% |
| Universities and colleges | — | — | 2.1 | — | — | 2.1 | 2.1% |
| Other Nonprofit institutions | — | — | 0.6 | — | 0.7 | 1.3 | 1.3% |
| TOTAL | $10.8 | $72.0 | $8.6 | $2.8 | $2.8 | $97.0 | 100.0% |
| Percent distribution, by performers | 11.1% | 74.2% | 8.9% | 2.9% | 2.9% | 100.0% | |

*Source:* Adopted from *National Patterns of Science and Technology Resources,* 1984, note 7.

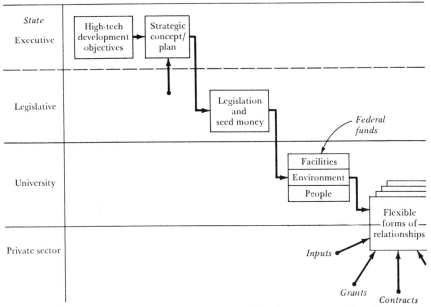

**Figure 5.3.** High technology development relationships (source: "Report on Fostering Emerging Technologies in Texas," Advisory Committee to the Texas Senate Committee on Business, Technology and Education, December 1984).

staff, and lastly innovation or incubator developments, middleman or supercenters, and research parks. Some of these are described below:

## CONSULTING
Consulting is basically an arrangement where individual faculty members enter into agreements with firms to provide services in their field of expertise. Universities generally allow faculty to spend up to 1 day per week in consulting activities, either on or off the campus. The needed linkages are thus minimal.

## RESEARCH GRANTS
Research grants are formal agreements or arrangements between universities and companies in which faculty members or groups of faculty, generally supported by graduate students, agree to perform research in a specific field, and on the university campus. The facility linkages are more extensive, but most likely include university-owned equipment and buildings.

## MAJOR CONTRACTS
Major contracts are a third form of university/industry linkages. In these cases, a university will enter into a multiyear, multimillion dollar contract

with a company to conduct research, generally in a broad area. For example, Monsanto has an agreement with Washington University in St. Louis to provide $33 million in biomedical research for which it will receive first right of refusal to develop any product under an exclusive licensing agreement.[10] At this point, the facility needs increase, and often require additional owned or leased space, again under university control.

## INDUSTRY COOPERATIVES
A fourth category is industry cooperatives, which occur when an entire industry sees a need for more basic research and more educated professionals and forms cooperative arrangements for working with universities. An example is the Semiconductor Industry Association, which spends millions annually in support of centers of excellence as well as individual research programs.

## CENTERS FOR ADVANCED TECHNOLOGY
Another type of linkage is illustrated by the New York State Centers for Advanced Technology formed in 1982. Seven centers are funded by the New York State Science and Technology Foundation, with matching grants from the private sector. The state funds provide for the purchase of equipment and support faculty, staff, and students. The centers are in place at Columbia University, Cornell University, Polytechnic Institute of New York, SUNY Buffalo, SUNY Stonybrook, Syracuse University, and the University of Rochester.[11]

## UNIVERSITY AND INDUSTRY CONSORTIA
Another linkage is university and industry consortia. In the past consortia were established to operate research facilities for the government. Today they can aid in the economic development of a region. These often become what are called supercenters, or middlemen, between universities and industry.

# MAJOR FACILITY TYPES

## Supercenters

Supercenters, or "buffer institutions," serve the purpose of meeting interdisciplinary needs, allowing for the sharing of resources, and provide flexibility in program development. This is illustrated in Figure 5-4, which outlines the purposes of industry-based research and of university-based research and shows how supercenters can serve as buffer institutions.

One example of supercenter is the Microelectronic Center of North Carolina (MCNC) located at the Research Triangle Park in North Carolina. The MCNC is a not-for-profit corporation formed in 1980 to in-

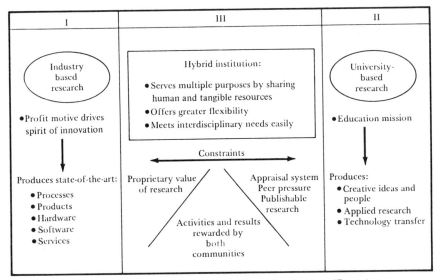

**Figure 5.4.** "Buffer institutions" as a research alternative (source: "Report on Fostering Emerging Technologies in Texas," Advisory committee to the Texas Senate Committee on Business, Technology and Education, December 1984).

crease research and educational resources available to the microelectronics industry and to encourage this industry to locate in the state. In 1981, the North Carolina General Assembly appropriated $24 million for construction and initial operations.[12] A second major example is the Microelectronics and Computer Technology Corporation (MCC). The MCC, owned by 12 U.S. microelectronic and computer companies, will be located on 20 a of land leased from the University of Texas, Austin Balcones Research Center. Another example is the Stanford University Center for Integrated Systems (CIS). The CIS is composed of Stanford University, the U.S. Department of Defense, and 20 private industry sponsors each of whom pledged funds toward the construction and operation of a new 90,000-sq-ft (8,361 sq. m) teaching and research facility. One representative will have the opportunity to work full time at CIS with the 73 affiliated Stanford faculty members.[13] At the level of consortia or supercenters, facility needs multiply; single-purpose buildings jointly sponsored by industry, government, and universities often are the result.

## Incubator Facilities

Innovation or incubator facilities are primarily those efforts by universities, university affiliates, private companies, or local economic development agencies to create physical space for emerging companies to locate temporarily. Because of its importance to economic development, the topic is covered in greater depth in Davidson's Chapter 6, which follows in this volume.

At this stage, offices are needed more than laboratories. The incubator centers generally offer support services in terms of product development, marketing, legal services, venture capital, and a whole host of related activities to encourage entrepreneurs in product or service development. As also shown in other chapters (6 and 15) a number of examples of incubator centers exist. One of the earliest is the University City Science Center adjacent to the University of Pennsylvania and Drexel University. It is estimated that one half of the 70 companies in the Science Center started as incubator industries there. Today the Science Center is composed of nine buildings on 19 a with a total employment of 4,500.

Another incubator center is the Yale-New Haven Science Park Development Corporation, which was formed in 1982 and is located on 80 a formerly housing the Olin Corporation. The site borders Yale University. Its tenants are primarily drawn from three groups: start-up companies for the Science Park New Enterprise Center, major corporation offices looking for space near downtown New Haven, and light industry that can use rehabilitated facilities.

Others include the Center for Entrepreneurial Development at Carnegie-Mellon University, the Incubator Program at Rensselaer Polytechnic Institute, the Advanced Technology Development Center at Georgia Institute of Technology, the Innovation Center at Ohio University, the Biotechnology Development Center at the University of Illinois-Chicago Circle, the Center for Innovative Technology in Virginia, and the Utah Innovation Center.[14]

The facilities vary in size from 25,000 sq ft (2,322 sq m) at Rensselaer to 1.2 million sq ft (111,480 sq m) at the Science Center (of which 400,000 sq ft [37,160 sq m] is leased to the Department of Health, Education and Welfare. One goal of the incubator facilities is to transfer the findings of basic research and development into new products and technologies. Many are spin-offs of academic research projects. As a result of the successes of these programs new proposals for similar projects have also occurred in Alabama, Louisiana, Mississippi, Tennessee, and North Carolina.

## Established Research Parks

As activity progresses from the theoretical to the product to the development stage, facility needs increase. To capture some of this activity university-affiliated research parks have developed. Historically these have been located at major institutions—Stanford, Princeton, the Research Triangle Park in North Carolina, Utah, and so forth. Further examples of university-affiliated research parks are given in Minshall's overview of high-technology parks in the previous chapter.

Each of these parks was started with a different purpose in mind. Stanford Research Park resulted in part from the research of a student named Russell Varian. His experiments led to the development of the Klystron tube, a new concept in microwave transmission. His desire to locate his plant near the university resulted in Varian and Associates becoming the first lessee of the Stanford Industrial Park in 1950. Today, the 655-a park has 80 tenants and approximately 26,000 employees.[15]

The North Carolina Research Triangle Park is a private corporation wholly controlled by the nonprofit Research Triangle Foundation, which includes the University of North Carolina, North Carolina State University, and Duke University. One of its objectives was regional economic development. A large number of the occupants of the park include governmental agencies such as the National Institute of Environmental Health Sciences and the National Environmental Research Center. The park, which was started in 1958, developed slowly. Today it has 45 primary tenants and 22,500 employees. In addition to research and production facilities, the North Carolina park includes a service center with a post office, banks, commercial space, rental office facilities, and a 200-room hotel.

The Princeton University Forrestal Center was started in 1972. Princeton's goals were to protect its research campus from encroaching residential and commercial development and, if it proved successful, to earn a return for the university's endowment fund. Today the 1,600-a Princeton Forrestal park has 23 primary tenants and employment of more than 4,000. This park also includes a hotel and conference facility, a commercial retail center, and 600 units of housing.

The University of Utah Research Park was started in 1965. This 310-a park now has 21 occupants and 3,000 employees. One area that is rapidly developing at the Utah park is in the area of biotechnology. As shown in Table 5-4, medical firms make up 28% of the park's occupants; another 10% are related to medical applications. Sixteen percent of the firms are engaged in energy research, and 10% in minerals. Firms whose speciality is either computer applications, electronics, or environmental sciences make up 20% of the firms.[16] It is also the home of the Utah Innovation Center, an incubator center.

While the Stanford, North Carolina, Princeton, and Utah parks have certainly been the most successful and are the giants of university-affiliated research parks, today a number of new parks are being started across the country.[17]

TABLE 5-4
**Distribution of firms, University of Utah Research Park**

| TYPE OF FIRM | % |
| --- | --- |
| Biomedical | 28 |
| Computer | 10 |
| Corporate office | 6 |
| Electronic | 4 |
| Energy | 16 |
| Environmental | 6 |
| Mechanical engineering | 6 |
| Medical | 10 |
| Minerals | 10 |
| Science/Research | 4 |
| TOTAL | 100% |

*Source:* Ira Fink and Associates, based on data provided by the University of Utah Research Park.

# New Research Parks

The following new research parks are under development:

• Arizona State University is developing a 320-a park in Tempe. The roads and infrastructure are in and sites have been leased for development.

• The Florida Research and Technology Campus, a project of the University of Florida Foundation in Gainesville, is developing 2,200 a of land. The Florida Progress Corporation, a private developer, purchased the first 190 a of land, with initial buildings now under way.

• The Ohio State University is in the process of developing two potential research parks totalling 200 a on lands adjacent to the campus in Columbus.

• Texas A&M University in College Station is developing a 434-a park. One unique aspect of this park is that it will seek to attract high-tech industries interested in agricultural and related areas of development.

• The University of Michigan, jointly with a private landowner, is undertaking the creation of the 820-a Ann Arbor Technology Park. Three parcels have already been sold on this site, which has common boundaries with the University of Michigan Botanical Gardens and is within 1 mile of the University of Michigan's north campus.

   The University of Missouri has selected the Trammell Crow Company as developer for a 225-a park on 700 a of land the University owns in Weldon Spring, near St. Louis. Ground was broken for this park, heavily supported by local and state sources, in 1986.

The University of Wisconsin–Madison is actively pursuing tenants for an office and research center, which will occupy a total of 130 a in a 320-a mixed-use development they are completing on a site 1 mile from the Madison campus.

- Finally, Washington State University in Pullman, is completing negotiations with a developer for the first project on its 158-a research park site. This 40,000-sq ft (3,716 sq m) research facility will include the university as a tenant in one half of the building.

## Research Park Characteristics

The university itself as an occupant of space in the research parks is becoming an important new characteristic of the parks. At the University of Utah, eight separate university departments have space in the park. At Washington State, the university will lease one half of the first building to be built in the park. At the University of Florida park, the university will guarantee the lease on one half of the first building. The reasons for this are straightforward. As universities obtain research grants, either from the government or from private sources, they are often short of space in which to conduct their research activities. Thus, like the developing company, they too seek space adjacent to campus space.

And many university research spin-offs, similar to private companies starting in innovation or incubator space, also have become business enterprises. For example, the Wharton Econometric Forecasting Associates, Inc., at the University of Pennsylvania is now a separate private company at the University City Science Center. The North Carolina Educational Computing Service, a unit of the University of North Carolina, has its principal offices at the Research Triangle.

## The University as Partner

At other parks, the university is a partner in the development of both the park and the companies. For example, at Utah the University holds equity positions in 24 companies.[18] One example of a successful project of this sort is the BSD Medical Corporation. In the late 1970s, a privately funded medical research project at the University of Utah Department of Bioengineering revealed that heat generated by radio frequency and microwave radiation reduced cancer tumors. This study formed the genesis for the corporation that today has emerged as a world leader in hyperthermia devices. The closest of BSD Medical's ties are with the radiation oncology researchers in the university's Department of Radiology. The university assists in clinically evaluating the products and defining product enhancements. Since 1978, BSD has had offices in the Utah Research Park.[19]

## Newer Facility Types in Research Parks

Some parks also include corporate headquarters offices, but most do not lease or allow other general office activities on the sites. Many of the park sites are occupied by agencies of the federal government. This is especially true of the North Carolina Research Triangle.

Finally the parks are developing into full multiuse activities including hotels, restaurants, banks and other support services. One important facility type located in the research park is the hotel/conference center. As noted earlier, at the Princeton Forestal Center there is a 300-room hotel/conference facility, known as the Scanticon Princeton Conference Center. At the University of Utah, construction is about to begin on a new $13-million conference facility. This 160,000-sq ft (14,864 sq m) facility will provide conference space and a restaurant, as well as lodging. This type of facility in conjunction with the university will provide space for the companies to hold meetings as well as to sponsor or attend regional and national conferences in their area of expertise.[20]

In most cases, the parks include multiuse sites that permit limited amounts of research applications, or in other words, manufacturing. Because some parks are being supported by their state economic development agencies or other agencies of the state, the initial start-up cost of the park is reduced, and these savings are passed on to park tenants.

## ECONOMIC DEVELOPMENT

## Regional Economic Development

As a result of these research park successes, universities are increasingly realizing that they can aid considerably in the economic development of their region. Since many of the industries that locate in research parks are local (that is, companies whose origins are in that area or region), the availability of a research park offers considerable incentive to local economic development.

Although there is not published national data on how many employment linkages actually occur, as shown in Table 5-5, at the University of Utah, about 20% of the Utah Research Park employees are graduates of the university.[21] Since a primary mechanism for technology transfer is through student recruiting, an industry that locates its facilities in proximity to a university obviously has a location advantage in recruitment.

TABLE 5-5
University graduates employed, University of Utah Research Park

|  | 1979 | 1982 |
|---|---|---|
| Total employment | 1,062 | 1,564 |
| University of Utah graduates | 206 | 300 |
| Percentage who are University of Utah graduates | 19% | 19% |

Source: Ira Fink and Associates, based on data provided by the University of Utah Research Park.

TABLE 5-6
University/industry linkages

|  | 1979 (%) | 1982 (%) |
|---|---|---|
| Use of faculty as consultants | 76 | 88 |
| University equipment used by industry | 88 | 71 |
| Industry equipment used by university | 59 | 61 |
| Participate in joint research proposals | 65 | 53 |
| Employees who are scientists or engineers | 46 | 37 |

Source: Ira Fink and Associates, based on data provided by the University of Utah Research Park.

Often the name of the university is one of the most important linkages between universities and industries locating in research parks. However, the linkages go beyond the names. As shown in Table 5-6, at the University of Utah Research Park, nearly nine out of 10 firms located in the park use faculty as consultants. Seven out of 10 of the firms use university equipment in their research. Six out of 10 have equipment that is used by the university. More than one half of the firms participate in joint research proposals.[22]

One final point about these linkages is also indicated in Table 5-6. Of the firms in the Utah Research Park, approximately 40% of the employees are classified as scientists or engineers. That means that 60% have non-scientific or administrative backgrounds. Thus, it is not only the university's presence that contributes to the success of the park but the availability of a skilled local labor force, often the result of an excellent community college or vocational school system in the area. This combination of factors has been one of the major reasons for the success of the Silicon Valley between Palo Alto and San Jose, and of Route 128 in Boston/Cambridge.

## Differences Between High-Tech and Low-Tech Firms

If one looks for a moment at the differences between high-tech firms and low-tech firms and their locational choices, it is easy to see why high-tech

**TABLE 5-7**
**Industry locational factors**

| HIGH-TECH COMPANIES | LOW-TECH COMPANIES |
|---|---|
| Higher skill | Access to markets |
| Tax climate | Access to raw materials |
| Academic environment | High volume transportation |
| Community attitudes | Low labor costs |

*Source:* Robert Premus, Note 5.

firms are attracted to the university setting. As shown in Table 5-7, low-tech firms need access to markets, raw materials, high-volume transportation networks, and low costs in terms of labor, energy, water, and waste treatment. High-tech firms require higher skilled employment, a favorable tax climate because of the entrepreneurial nature of many of their ventures, the academic environment that allows exchanges with staff and faculty, favorable community attitudes, and a whole series of aspects that fall under the umbrella of increased quality of life for their employees.[23]

# SUMMARY

In terms of facility linkages, one factor today is *academic capitalism*—in other words, universities assisting in the commercialization of academic discoveries or assisting private companies.[24] These are instances of universities not being actively involved in the product development but expecting to share in profits and returns.

This activity extends from assistance to individual research or to groups of researchers or entrepreneurs to use of resources, including university equipment, facilities, land, and endowment or investment capital. Most serve the goal of technology transfer. In the purest sense it is done in support of education and the free pursuit of knowledge. At each level of development the facility and equipment requirements vary.

While there are strong points in high-tech linkages, there are also concerns and conflicts over these activities. Some include scientific communication and proprietary rights, exclusivity of patents, conflicts of interest and commitment, and the use of university facilities and instrumentation.

In general, most linkages between universities and industries can be of benefit to both if certain conditions are met. One of these is that the university is free to choose which programs to accept and which to reject. A second is that the linkages between the company and the university are based on educational programs of intrinsic value to both.

Clearly, university and industry facilities can foster high-tech linkages. One of the most important functions the facilities serve is that they aid in technology transfer by placing the companies in a better position for student recruiting. The students, like the companies, often prefer to remain in the local area. This is often an overlooked aspect of facilities in fostering high-tech linkages between universities and industries. Chapter 7, by Franklin Becker, throws light on detailed aspects of facility management strategies and organizational culture that govern decisions and resources for those facilities most responsible for effective university/industry linkages, namely university research laboratories; and particularly on the interdependence between the quality and caliber of faculty and students (hence of research, and ultimately of research spin-offs leading to local economic development), on the one hand, and the quality of laboratory facilities and equipment, on the other.

## CONCLUSIONS

From this perspective about high-tech linkages a number of conclusions can be drawn:

1. In terms of research linkages, universities, industries, and the government are interdependent.[25]

2. The linkages are complicated.

3. The motivations of universities, industry, and government are different and the differences are basic.[26]

4. Successful linkages require careful planning and years to execute.

5. Much of the result of the linkages is serendipitous.

6. Linkages require not only mutual respect but mutual interest.[27]

7. In the past there was a time lag between scientific advance, new technology, and industrial development. As a result of the loss of time lag, the filters that were in place have disappeared, and a radical change is occurring in relations between universities and the high-tech sector.[28]

8. Significant increases in industry funds for universities will be required to support research activities. However, industry cannot compensate substantially for the erosion of federal support.

9. A new type of facility, the middleman, or supercenter, will continue to be developed.[29]

10. The small companies will be inclined to develop new ideas. But brilliant science alone will not be enough for success.[30]

11. As the ties between universities and industries grow closer, a balance must be struck between the intellectual independence of the scientist and the university and the industrial desire for commercial support.[31]

12. Within universities there will remain two conflicting groups: the gatekeepers of traditional values and the entrepreneurs.

Overall, the development of university and university-affiliated facilities will continue to play an important role in providing the links within which technology transfer and academic capitalism involving universities can take place. The locational advantage of sites near universities offers more opportunities for the linkages to occur.

## NOTES

1. Thomas W. Langfitt, "Epilogue," in *Partners in the Research Enterprise: University–Corporate Relations in Science and Technology*, ed. Thomas W. Langfitt et al. (Philadelphia: University of Pennsylvania Press, 1983), p. 175.
2. Bryce Douglas, "Rapporteurs," *Partners (op. cit.)*, p. 157.
3. Thomas D. Kiley, "Licensing Revenue for Universities: Impediments and Possibilities," *Partners (op. cit.)*, p. 65.
4. Neil H. Brodsky, Harold G. Kaufman, and John D. Tooker. *University/Industry Cooperation: A Preliminary Analysis of Existing Mechanisms and Their Relationship to the Innovation Process* (New York: New York University, Graduate School of Public Administration, Center for Science and Technology Policy, June 1980), p. 9.
5. Robert Premus, "The Research Park Phenomenon," presented at the Urban Land Institute Conference: "The University/Real Estate Connection," Palo Alto, Cal., June 11, 1984.
6. Robert Premus, "Location of High Technology Firms and Regional Economic Development," a staff study prepared for the use of the Subcommittee on Monetary and Fiscal Policy of the Joint Economic Committee, Congress of the United States (Washington, D.C.: U.S. Government Printing Office, 1984), p. 22.

7. *National Patterns of Science and Technology Resources, 1984,* National Science Foundation, NSF84-311, (Washington, D.C.: U.S. Government Printing Office, 1984), p. 10.

8. Frank Press, "Core Technologies in the National Economy," *Partners (op. cit.),* p. 42.

9. George M. Lowe, "The Organization of Industrial Relationships in Universities," *Partners (op. cit.),* pp. 71–73.

10. Samuel B. Guze, "The Monsanto–Washington University Biomedical Research Agreement," *Partners (op. cit.),* pp. 53–58.

11. "New York State Centers for Advanced Technology," brochure (Albany: New York State Science and Technology Foundation, 1983).

12. Ira Fink, "The University as a Land Developer," *Planning for Higher Education,* published by the Society for College and University Planning, vol. 12, no. 1 (Fall 1983), p. 28.

13. *Facilities Planning News,* vol. 3, no. 2 (April 1984) (Orinda, Cal.: Tradeline, Inc.), p. 1.

14. Mihailo Temali and Candance Campbell, *Business Incubator Profiles: A National Survey,* (Minneapolis: University of Minnesota, Hubert H. Humphrey Institute of Public Affairs, July 1984), pp. 87–102.

15. Fink, *(op. cit.),* p. 26.

16. Charles Evans, University of Utah Research Park, February 1985.

17. Ira Fink, "The New University R&D Parks," *Facilities Planning News,* vol. 3, no. 6, (December 1984) (Orinda, Cal.: Tradeline, Inc.), p. 1.

18. *Research Park Report,* Winter 1985 (Salt Lake City: University of Utah, Research Park), p. 1.

19. *Ibid.,* p. 2.

20. *Ibid.,* p. 1.

21. Evans, *(op. cit.).*

22. Evans, *(op. cit.).*

23. Premus, "The Research Park Phenomenon," *(op. cit.).*

24. "Report on Fostering Emerging Technologies in Texas," Advisory Committee to the Texas Senate Committee on Business, Technology and Education. (Austin, Texas: December 1984).

25. Thomas D. Kiley, "Licensing Revenue for Universities: Impediments and Possibilities," *Partners (op. cit.),* p. 59.

26. George M. Lowe, "The Organization of Industrial Relationships in Universities," *Partners (op. cit.),* p. 69.

27. Peter Barton Hutt, *Partners (op. cit.),* p. 81.

28. George E. Palade, *Partners (op. cit.),* p. 83.

29. Elbert Gore, Jr., "Recombinated Institutions: The Changing University–Corporate Relationship," *Partners (op. cit.),* p. 127.

30. Ronald E. Cape, *Partners (op. cit.),* p. 117.

31. Gore, "Recombinated Institutions," *Partners (op. cit.),* p. 125.

CHAPTER SIX

# THE BUSINESS AND INDUSTRIAL INCUBATOR—A TOOL FOR LOCAL DEVELOPMENT AND ENTREPRENEURSHIP

## COLIN H. DAVIDSON

**A**n incubator can be defined as: "An organization which helps with the development of new enterprises, by providing them with space, services and counselling."[1] And it can also be defined as follows:

> *a local grouping of young, in some cases newly founded, firms and undertakings:*
> * *whose business centers on the development and marketing of new-technology products and processes,*
> * *whose advisory needs are high,*
> * *and which, in many cases, need a large amount of capital during the start-up phase but can offer no adequate security;*

The group ensures the

> * *provision of common office and service facilities;*
> * *availability of a wide range of services for common use;*

- *provision of advice, information and contact brokerage by a qualified center management;*
[on a]
- *local to regional base.*[2]

These definitions contain essential information for understanding what an incubator is, and what its purpose is. First and foremost, an incubator is an *organization* set up for the specific purpose of nurturing young enterprises as they grow into the world of business or industry, of markets, profits, and losses. Its tools include a *building* (its "hardware") and *advice* (its "software"); it is a balanced combination of the two that enables each incubator to fill its mission *in its particular context of local and regional economic development.*

In Section I of this chapter, the subject is approached from the point of view of local and regional economic development, because long-range planning information must be assessed and strategic decisions made prior to implanting any incubator in a community. In Section II, the architecture of the incubator is discussed, because novel requirements guide the design process whether an incubator is to be housed in a new or in a rehabilitated facility.

## SECTION I: THE INCUBATOR REGIONAL ECONOMIC DEVELOPMENT

### The Context

In the late 1980s, the political and economic discourse in most industrialized countries is being driven by economic problems associated with job losses and new processes of job creation. It is well known that the older "heavy" industries are *negative* sources of employment, as their labor-intensive operations are moved to cheap-labor countries, or as robotized operations begin to predominate; it is now equally well known that the highest performing enterprises, in terms of innovation and job creation, are small businesses. For example, a 1981 study showed that about two thirds of jobs created in the United States during the survey period were created by firms with fewer than 20 employees.[3] Small businesses are stated to be 24 times more innovative than large corporations.[4] At the same time, *high tech* is seen as the substitute field of endeavor—with start-up companies showing the most promise, *provided they survive the difficulties of managing their own growth.*

However, setting up and developing new enterprises (whether in the service or manufacturing field) involve more than political or scientific well-wishing; problems of social organization and resources management must be resolved, and a technical "culture" developed. Depending on the sociopolitical context, the development of a fertile industrial or commercial milieu will be the result of more or less planned decision making; in all cases, various skills (technical inventiveness, management ability, financial know-how, market flair, etc.) must be brought together around the entrepreneurs and around their innovative ideas. Good working and living conditions are needed to attract the entrepreneurs of tomorrow and their staffs; these conditions can be attained only by coordinated decision making in the fields of regional planning and architectural design. The impetus for a systematic approach to planning and design is often provided by the wish to set up an incubator in a community.

The contemporary context contains additional complexities that are favorable to the establishment of an incubator; they are related to the current worldwide climate of instability and change. Large bureaucratic organizational structures are unable to adapt to them, leading, for example, to the adoption—within the largest corporations or bureaucracies—of project groups and the task-organization approach to management, tantamount to setting up "firms within firms." Small businesses, however, are able to fit into this situation of instability and change by filling niches that large corporations, with their inflexible structures, cannot readily occupy. Job enrichment—another contemporary concern—also works in the same direction, if for different reasons. In the largest enterprises, previously fractioned tasks are now deliberately reassembled and responsibility delegated; again, small businesses offer an "ideal" model, since their small workforces can participate broadly in a number of integrated operations. Furthermore, decentralized small enterprises can regroup in dynamic networks for specific missions, adopting as many forms of cooperation as there are partners. Today, "small is beautiful" assumes a novel significance in the world of management.

In this context, scientific and technical knowledge is necessary but not sufficient; survival of the small firm requires a balanced and systematic approach often lacking in their founders, particularly during the start-up phases. Haude notes that young enterprises are known to have certain types of problems, such as:

— *financing problems and difficulties in preparing documents required by banks for the approval of loans;*
— *administrative and legal questions, particularly in relation to registration as a business . . . ;*
— *preparation of technical product documentation and appraisals for the presentation or approval of a product or for comparing products, using an approved and customary format . . . ;*
— *marketing problems, where clients behave otherwise than expected;*
— *commercial and organizational problems, especially future-oriented cost accounting and export management;*

> — *standardizing design and calculations [in line] with [the]*
> *preparation of verifiable documentation;*
> — *materials testing;*
> — *product trials and systematic evaluation of results.*[5]

An incubator, with its components of appropriate space and oriented advice, provides the resources and the setting in which these problems can be overcome or even bypassed.

## Typology

The business or industrial incubator is, therefore, a new concept, explicitly designed to help young enterprises during their start-up years. An incubator can be classified in terms of its sponsors and their objectives: profit, job creation, technology transfer, economic development or diversification, tax base enhancement, prestige/image making, provision of investment opportunities, etc.[6] An incubator can, therefore, be any of the following:

- Public, private, mixed

- University based, corporation sponsored, municipal

- For profit, nonprofit

- Sectorial, nonspecialized

- High tech, socially oriented (for example, for minorities)

- Urban, rural

Four main components characterize all incubators:

1. A building in which new or existing firms can find space for rent at or below market rates

2. Shared services

3. Access to counseling on management, marketing, accounting, research and development, technology transfer, exporting, etc.

4. Access to seed capital (to be distinguished from risk capital)

The importance given to each of these components will vary from case to case; indeed, the variations can be so great that even incubators "without walls" have been proposed, where all the services and counseling are provided, but without any rental space* (the converse—flexible rental spaces without services—also exists, but is more properly referred to as an "industrial motel").

The incubator is, therefore, an organization in which young businesses find services and space to rent. The services are designed to complete the needs of the firms, even to complement their shortcomings; they include secretariat/word processing, reception/private branch exchange, photocopy/printing, receiving/shipping, and microcomputing (often included fully or partly in the rent). In addition, ready access to consultants is assured at cost† (often retired people in the local community, who advise on bookkeeping, management planning, preparation of business plans, etc.). Table 6-1, based on data provided by King,[7] gives a list of the services presently available or requested; Table 6-2 indicates additional resources that are useful for start-up firms. For further convenience, the rental space (at least according to some sources) should be, and often is, offered at below-market rates; however, it should be realized that the very existence of common services reduces the necessary space for each tenant by so much that even at market rates total rent will be far less than elsewhere.

In this view of the incubator, it is clear that the building is a means and not an end. Obviously it must provide spaces for the shared services and rooms for consultations with the counselors. As well as the rental spaces, it includes meeting rooms of different capacities, laboratories, photo labs, workshops for prototyping, etc., available to the tenant enterprises on a reservation basis (again, charging policy varies from case to case; sometimes there is no charge, sometimes an at-cost charge after a certain amount of usage by a particular tenant).

Within the context of technological change, the concept of the incubator is seen as a way of firmly establishing high-tech enterprises founded by a person holding an original technological idea and a vision of its practical application. While it is obvious that the young *scientific* enterprise needs helping—through provision of services and of advice—it has now become evident that within the same context of rapid change, *any* young enterprise can profit from similar support. This observation has given rise to the idea that the incubator approach should not be limited a priori to the fields of advanced science and high technology; it can be applied equally well to other socioeconomic situations. The concept of the "socially oriented" incubator has emerged, for example, for job creation in economically weak

---

* Incubators *without* walls exist, for example, at a university where the emphasis is on providing the *services* rather than the space (obviously there will be space for the services—possibly in existing university buildings); the presumption then is that the incubated firms know how to and do find their own space without requiring costly investments by the sponsor.

† Sometimes the consultants agree to provide a certain number of hours advice at no charge, regarding this as a marketing investment in developing their own clientele.

**TABLE 6-1**
**Shared services in incubators.**

| SERVICES MOST FREQUENTLY PROVIDED[a] | | SERVICES MOST FREQUENTLY ASKED FOR[a] | |
|---|---|---|---|
| *PUBLIC SPONSOR* | *PRIVATE SPONSOR* | *PUBLIC SPONSOR* | *PRIVATE SPONSOR* |
| Meeting rooms | Photocopying | Photocopying | Photocopying |
| Maintenance | Secretariat | Meeting rooms | Receptionist |
| Shipping | Maintenance | Receptionist | Secretariat |
| Mail service | Assistance with | Word processing | Assistance with |
| Photocopying | business plans | Secretariat | business plans |
| Furniture rental | Advice about grants | Shipping | Financing debt, capital |
| Secretariat | Meeting rooms | Security service | Accounting |
| Word processing | Word processing | Publicity, marketing | Advice about grants |
| Receptionist | Publicity, marketing | Accounting | Publicity, marketing |
| Cafeteria | After-incubation | Courrier, 24-hr | Legal services |
| Assistance with | rental arrange- | telephone | Telephone answering |
| business plans | ments | Assistance with | Book keeping |
| Computers | R & D assistance | business plans | R & D assistance |
| Library | Training programs | Financing debt, | Maintenance |
| Audio visual | Mail service | capital | Word processing |
| equipment | Security service | Telex | |
| Security service | Financing debt, | Library | |
| | capital | Furniture rental | |

*Source:* King, et al., note 7.
Note: Similar information is provided, in more detail, in David N. Allen, *Small Business Incubators and Enterprise Development,* (Carlisle, Pa: National Business Incubation Association) Tables 5.3.1 to 5.3.3.
[a] Services are listed in order of decreasing frequency of provision or request.

**TABLE 6-2**
**Additional services useful for start-up enterprises**

Free access to libraries containing
• Up-to-date technical journals in a variety of fields
• Updated collections of "short life" (unpublished) reports
• Collections of standards and guidelines
• Collections of monographs and audio-visuals

Selective supply and preparation of retrospective documentation and current awareness

Establishment of personal contacts with colleges and R & D institutions.

Supply of manpower
• students acquiring practical experience
• students conducting research in related areas
• graduates with appropriate research skills

Access to computing facilities and to telecommunications networks

*Source:* Haude, note 5.

neighborhoods. However, there is disagreement (1) as to whether the incubator approach is better limited to the needs of high-tech enterprises or, if not, (2) as to whether high and low tech can be mixed in the same incubator. On the first point, the needs of *all* young enterprises are virtually similar (see, for example, the list quoted from Haude, above), irrespective of the technological niche they occupy, implying that they can *all* benefit from the resources of an incubator. On the second point, a special responsibility falls on the incubator designer to separate the various types of enterprises where necessary (e.g., where noise and dust may pose problems), without creating two "classes" of tenants (see, for example, the second design presented in Part II of this chapter). Furthermore, all the tenants require easy access to appropriate outside resources; in cases where the tenants are very different, the appropriate level of proximity to widely differing resources may raise difficulties, so that in effect the kinds of firms to be housed together are likely to be predetermined by this consideration above all else. As Naylor explains:

*"High technology" businesses clearly have specific features that make them different in some respects from what might be called "low" and "no" technology businesses. Specifically, many high tech businesses need access to support services able to provide them with access to technical expertise and sources of information. A need for skilled personnel may make it advantageous to site such companies in . . . innovation centers attached to higher education establishments.*

*However, while high tech businesses can differ in some ways from other businesses, in* commercial *terms they have the same requirements. A need for premises, finance, management and a base technical expertise exists.*[8]

Because of the wide variety of types, a proper understanding of what an incubator is starts from the generic fact that an incubator is an element of small business development, and not just a building sheltering several tenant enterprises. Seen in this way, the incubator is a tool for economic development and business growth; to be successful, it must be integrated into a plan for local economic development, and it must be backed up with strong local support.

## Local Economic Development

Incubators, as has been pointed out, correspond to a wide range of variants, depending on the circumstances leading to their development. Some are established in old industrial neighborhoods in need of revitalization, whereas others are near universities or financial and business districts.

A successful incubator is rooted in a dynamic community whose priorities and needs it reflects. Depending on its objectives, an incubator may include: social facilities open to the community or to employees of tenant firms (such as a nursery school or restaurant), scientific equipment (specialized labs with sterile hoods, oxygen supplies, computer networks, etc.), workshops to prepare prototypes and documentation (metal and plastics shops, silk screen and printing, etc.). Sometimes, an incubator is included in a science or industrial park, which allows "graduating" firms to make an easier transfer to their own facilities in the park, possibly being able to negotiate continued access to services in the incubator or nearby.

An incubator seems to contribute significantly to the chances of survival of the young enterprises it serves; indeed, available figures seem to indicate a switch from a high failure rate to high probabilities of survival (failure rates are reduced from 50 to 14% after 5 yr).[9] Most important, the incubator is a tool—to be considered among others—for the strategic development of the community involved. By sponsoring promising enterprises, the incubator provides opportunities for many investors. This may take the form of investments by large corporations, which see in the incubated enterprises a potential for profitable transactions. Local investors may also participate in the risks of the new firm, in exchange for a share of profits once the threshold of profitability has been reached. And risk capitalists may intervene, through the mechanisms of participation in share capital (though this form of investment more often occurs after the enterprise has left the incubator, and is progressing through the next stage of its development).

## Planning

Planning an incubator is an operation requiring care and foresight; it falls into the category of "strategic" regional or local planning, and it requires knowledge about organizational design. The broad objective of the planning process is to obtain the best benefit from the resources and potentials of a given environment (scientific, technological, social, economic). Even if the kinds of services and counseling required by the enterprises can be grouped into a few broad categories (as shown in Table 6-1 for example), the most appropriate response will vary from region to region, and from case to case. The incubator, if it is successful, will act as a focal point for a community (a university, a municipality, or a neighborhood, for example), but its repercussions will spread far beyond its walls. Indeed, in some cases, government granting agencies condition their offers of financial support for an incubator to a requirement that the services and counseling be made available to enterprises that are *not* part of the incubator, thus defusing potential claims that conditions of unfair competition are being encouraged with public money.

For an incubator to be established, it is necessary to have the following:

- A "market" of potential entrepreneurs
- A project "champion" or project leader
- A small, committed support and organizing group
- Active support by the community
- Financing adapted to the chosen approach

A possible approach to the decision-making process is outlined in Figure 6-1. The process properly starts with a systematic statement of objectives. This statement of objectives should not be "descriptive" (e.g., "to have an incubator near our university"), but rather "prescriptive" (e.g., "to favor the creation of high-tech enterprises by our graduates"). The statement of objectives is followed by a study of environmental factors and of the potentialities of the sponsoring organization. Possible solutions are developed, tested in terms of potentialities and environmental factors, and finally matched against objectives. The most satisfactory solution is identified, tested, and then developed through to implementation.

## Finance and Management

The part played by private or public finance varies considerably, depending on the regional social and economic context; it is dangerous to generalize. In the United States, a large number of private incubators exist (though, in fact many—if not the majority—of them receive grant support in one form or another), whereas in Canada and Europe, public financing seems to predominate more explicitly. The main variable is the nature of the sponsor and partners, and their capacity to raise the necessary financing from private or public sources. A number of public grant programs can be exploited, such as make-work programs, grants for rehabilitating abandoned buildings, and regional development subsidies. Public or semipublic incubators are often not expected to show profit, since they are viewed as investments in community development.

In all cases, financing the *incubator* must be distinguished from financing the young *enterprises*, because the sources and conditions of financing are fundamentally different in each case. A further distinction can also be made; for example, in some incubators, the services are provided by a non-profit corporation, whereas the buildings are managed for profit; others, as has been mentioned, are totally profit oriented or non–profit oriented.

From the management point of view, an "average" incubator will probably house from 20 to 30 enterprises, whose stay lasts between 18 months

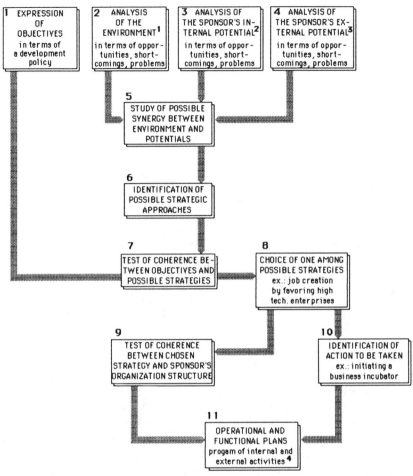

**Figure 6-1.** Strategy for economic development: decision sequence. Copyright ©
ACTE (Canada), Inc.
**Notes:** (1) ENVIRONMENT: elements outside the sponsor's organization which he
cannot control, but which influence him; (2) INTERNAL POTENTIAL: elements within
the sponsor's organization which he can control; (3) EXTERNAL POTENTIAL:
elements outside the sponsor's organization which he can control and/or influence; (4)
INTERNAL AND EXTERNAL ACTIVITIES: activities within the sponsor's organization
or outside it.

and 3 (sometimes 5) years. Though the incubator staff is small (usually
about 10 people, plus the counselors), management can be quite complex,
and must be individualized to suit the tenants, since each enterprise is
growing rapidly, changing its requirements constantly, and often facing
urgencies and crises as the development process evolves.

## The Enterprises

New enterprises fall into one of four classes:

- Soft companies
- Hard companies
- Garage companies
- Spin-out[10]

*Soft* companies are based on individual contracts and development work; *hard* companies tend to offer a standard range of products to an "anonymous" market; *garage* companies are similar to hard companies but started out by developing an innovative product, with financing on a largely personal basis; and *spin-out* companies are those that are founded by someone who had previously worked in a large company on similar products or concepts, for which most of the research and development have already been done.

Whatever their type, identifying new enterprises cannot be left to chance; one of the best sources of information lies with local groups such as financial institutions, accountants, lawyers, marketing and management counselors, education and research institutions, branches of government agencies, local governments, chambers of commerce and other professional associations, investors, and risk capitalists. The actual search for "candidate" enterprises must be oriented to reflect the objectives of the incubator: specialized/unspecialized, high-tech/socially oriented, limited to new enterprises/open to existing companies, restricted to certain types of enterprises/unrestricted, etc. Also, at this stage, a policy should be established regarding length of stay in the incubator: Is it limited—e.g., to a certain predetermined period or until the firm attains a given level of financial and commercial stability—or is there no limit?

Once identified, the process of selecting candidate entrepreneurs usually involves four steps (though sometimes these steps are not distinguished very clearly).

1. Candidates are encouraged to develop their original ideas; often a system for "foster parenting" potential ideas is set up, where one or two monitors explain to the candidate that setting up an enterprise, though difficult, is often more plausible than might appear, and suggest appropriate action to be initiated; often, outline business plans are prepared during this stage.

2. The process of *technical* evaluation then takes place, with each candidate explaining his or her ideas to a panel of technical advisers; these advisers are chosen for their expertise in the appropriate area of science, technology, or commerce, and are bound by the ethics of secrecy (candidates often approach such a meeting by being reticent about "revealing" their good ideas; however, they soon realize that they have probably more to gain from the panel than they thought at first, and experience shows that these evaluation meetings usually lead to a considerable improvement in the original idea).

3. A business-oriented panel meets with candidates to assess their outline business plans for feasibility, and to discuss how to finance the likely growth of the enterprise.

4. If the outcome of all these steps is positive, the incubator management team checks the candidates' references and negotiates their entry into the incubator, and draws up the lease.

Leases are generally flexible, and rents may be progressive (related to growth) or at a fixed rate (on a per-footage basis). Initially, the individually rented areas may be very small (for example, enough private space for a table, chair, and filing cabinet, and a post office box, with right of access to meeting rooms, workshops, or labs), increasing to 200, 300, or 1000 sq ft (approximately 20, 30, or 100 sq m) after 6 or 12 months of activity. As well as determining how much space is to be rented, and how changes in requirements are to be dealt with, the lease probably also covers terms of access (1) to the basic services (see Table 6-1) and to the meeting rooms, workshops, labs, etc., (often included in the rent with expenses counted at cost) and (2) to meetings with the counselors (nominal fee scales are agreed upon, and a given number of "free" consultations stipulated). The lease may include conditions governing access to outside resources, such as nearby university computers or specialized libraries (see Table 6-2). The length of the lease and policies regarding renewal are also specified. In many cases, leases (of a few months' duration) cannot be renewed beyond 18 months or 3 years, though a limit of 5 years is also encountered.*

---

* It is sometimes argued that there should be no time limit, since the firms would otherwise be obliged to incur the disruption and expenses of moving at what is probably a critical stage of their growth. In these cases, the incubator sponsors have a policy of financing and constructing further space as the need arises. It has also been found that firms may well leave the incubator of their own volition before they are obliged to; for example, some tenants feel that they have reached a time when the label of "incubator tenant"—however useful it may have been at the outset—begins to have a negative impact on their customers.

These somewhat elaborate selection and rental procedures are needed (1) to ensure that the strategic objectives of the incubator are satisfied, and that only appropriate firms benefit from the organization and its assistance, and (2) to protect the reputation of the incubator and its tenants, and thus to "guarantee" the quality of the tenant enterprises as they look for help outside the incubator. Lenders or investors are influenced by this guarantee and are more ready to entertain the idea of helping enterprises that have qualified for admittance to a reputable incubator, precisely because it is known that they have been evaluated systematically from the technical and commercial points of view, by the best available experts.

## Incubator Growth

Incubators, like the small enterprises they house, are also small businesses in their own right, moving through various growth phases themselves. Allen and Hendrickson-Smith[11] explain that there are "three stages of incubator development: (1) start-up, (2) business development, and (3) maturity, which on average, span a 4- to 6-year period." During these stages, the priorities for the incubator management will change. At start-up (i.e., from the first launching of the concept through to the break-even level of occupancy), problems of "real estate development" will predominate; financing the incubator itself will determine the rate of development. Here, local or community support is particularly important. Once the business development stage is reached, the incubator management team is able to fully devote its time to the tenant enterprises—as, indeed, was expected of it all along. It is at this stage that the consultancy services reach their proper levels, so that community involvement shifts from real estate financing to sponsoring entrepreneurial success. The maturity stage is reached when space in the incubator is "overbooked," so that expansion has to be envisaged. By then, a "graduation" policy will have been properly implemented—even if, at the earlier stages, there was a strong temptation to renew tenants' leases beyond the reasonable limit, if only to keep up the level of rental income.

## SECTION II: THE ARCHITECTURE OF THE INCUBATOR

## Design Criteria

Whether located in a recycled or in a purpose-designed building, programming and planning start from decisions about the shared services, which determine above all else the particular character of each and every incubator. For it is the judicious mix of *private* and of *common* spaces that allows the necessary independence for each firm to carry out the demanding tasks associated with its new commercial or industrial venture, while opening the way for the "accidental" meetings that can generate new ideas and new business. Indeed, experience shows that meetings between entrepreneurs can lead beyond the exchange of friendly advice, to the formation of joint ventures or the negotiation of subcontracts.

As for the private rental spaces, they must be flexible to cope with the needs of the different enterprises and, above all, with their different rates of growth. However, this requirement for *flexibility* runs counter to the requirements of *separation*—whether as protection from noise, vibrations and other forms of pollution, or as a bulwark against eavesdropping.

At the *feasibility* phase of the design process, it is important that the role of the intended incubator be made explicit in its context of regional and local economic development, in order to assess (1) the kinds of enterprises that are likely to become tenants and (2) the likely rate and duration of occupancy; many design decisions depend on this knowledge, including deciding on how to phase the construction work, since it is extremely damaging, at any phase, to be in the situation where a few firms are "lost" in large untenanted spaces. Approaches to and sources of funding must also be made explicit, because of the particular constraints and conditions that are attached to most of them. In this way, a complete and dynamic budget picture can be built up and matched against the original objectives and long-term plans. Obviously, zoning, infrastructure, and other urban planning questions must be fully ventilated then.

At the *functional programming* phase, information is analyzed about (1) the nature of the services and shared facilities to be provided, (2) the amount of counseling expected, and (3) the likely activities of the future tenant enterprises. However, it is extremely important to realize how much these activities may change over a period of time, particularly regarding the tenants who, by their nature, are constantly evolving; in a sense, it is necessary to program for *ranges* of requirements with critical thresholds, rather than to seek for definitive statements of needs. Gross and net areas will be decided upon.

At the *performance criteria* phase, this information must be translated into lists of human and technical requirements. Human requirements should include indications about how to ensure a high quality of life at work (a subject likely to be overlooked by the entrepreneurs, since they are probably more concerned about making a success of their firm than they are about their employees' or their own physical and mental well-being). Human workplace requirements also cover both the "measurable" physiological aspects (heat, light, and sound) and the intangible psychological aspects (view, color, aesthetics, etc.). Technical requirements, as well as including the obvious needs of separation versus flexibility and adaptability (acoustical and vibration separation and privacy on the one hand versus scope for rapid change with minimum disturbance on the other), also cover statements about (1) particular supplies (oxygen, natural gas, sterilized ventilation, distilled water, high-capacity electronic hookups, etc.) and (2) specially equipped spaces (scientific laboratories, dark rooms, etc.). Structural criteria, such as floor loadings, spans, and clear heights, will be fixed.

## Procurement

At the *procurement* phase, whether this comes before design (as in a design/build situation) or after (as with design/bid/build), it is particularly important to make sure that the expected levels of functional performance and quality are properly expressed (in performance specifications for design/build, in descriptive drawings and specifications for design/bid/build). Mechanisms should be set up at this phase to ensure their respect. In other words, choosing the principal members of the building team (architect, mechanical engineer, and main contractor) must be based on an assessment of their ability to perform in novel situations; like other procurement decisions, choosing the team should form part of the strategic plan for implementing the incubator development policy, and should be decided upon early.

## Some Examples

Table 6-3 lists, in the form of ranges of values, some of the main characteristics of incubators recently completed in the United States and Canada; Figure 6-2 illustrates four recent incubator projects.

The incubator shown in Figure 6-2a (never built) was to be located in a recycled nineteenth-century industrial building, with a new annex at the rear containing laboratories, photo lab, and workshop. The project includes community facilities (cafeteria and nursery), considered as fledgling enterprises in their own right, and a multipurpose room, all open to the public. "Spontaneous" meetings are encouraged by enlarging—on each upper floor—the main vertical circulation areas (featuring bay windows) to allow for sitting space and coffee machines. About 50 enterprises could be accommodated. Areas break down as follows:

- Rental (including cafeteria and nursery): 48%

- Services and shared spaces (including offices, meeting rooms, multipurpose room, shipping, toilets, laboratories, workshops, sitting areas, etc.): 29%

- Circulation (including part of the halls, stairs, elevators, etc.): 23%

The incubator shown in Figure 6-2b was also designed for a recycled factory, located between a busy street and a canal that has recently been

**TABLE 6-3**
**Data about buildings housing incubators**

| | Incubator sponsor | | | |
|---|---|---|---|---|
| | Public (13)[a] | Nonprofit (8) | University (7) | Private (22) |
| Rent (U.S.$/sq ft/yr)[b] | <0.50–4.00 | <1.25–3.00 | 6.00–15.00 | 1.5–16.00 |
| Min/Max total areas (sq ft) | 1,000,000 | 600,000 | 1,200,000 | 400,000 |
| | 20,000 | 15,000 | 25,000 | 20,000 |
| Types of tenants (%) | | | | |
| Office | 31 | 63 | 43 | 100 |
| High tech | 8 | 25 | 100 | 100 |
| Manufacturing | 100 | 88 | 43 | 100 |
| Retailing | 15 | 38 | — | — |
| Warehousing | 31 | — | — | — |

| | Range of percentages | |
|---|---|---|
| Rental[c] | 80%–48% | |
| Services + shared space[c] | 29%–5% | often about 25% |
| Circulation[c] | 37%–9% | |

*Sources:* Temali and Campbell (note 9) and survey by the author.
[a] No. surveyed.
[b] $ 1984.
[c] These figures are based on a smaller sample; also there is the risk that the data are based on differing interpretations of the categories by the respondents.

"upgraded" and turned into a linear urban park and cycle track. The principal design problem stemmed from the great depth of the existing building, and the difficulty of letting daylight into the central areas. The proposal is based on the concept of a pedestrian mall running the length of the building on each floor, slightly off-center (nearer the canal), lit by skylights and landscaped light wells. On the noisier (deeper) side of this mall, light manufacturing enterprises are located, and —in the smaller, quieter spaces on the other side—space for the "brainpower" firms. This mall widens out on the canal side at each upper floor to allow for a sitting area and coffee machines. The pedestrian mall assumes its full importance in this project because each enterprise has a "shop window" opening onto it for displaying its activities or products; this continuous run of shop windows also houses ducts and specialized service lines in its upper part, easily accessible for adaptations. The project allows for between 55 and 65 enterprises to be accommodated. Areas break down as follows:

- Rental (including restaurant, nursery, gym, etc.): 50%

- Services and shared spaces (including offices, meeting rooms, shipping, toilets, laboratories, workshops, sitting areas, etc.): 13%

- Circulation (including pedestrian malls, stairs, elevators, etc.): 37%

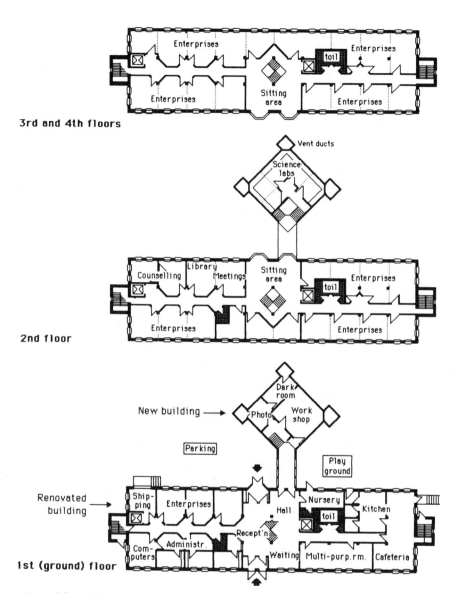

**Figure 6-2.** (a) Project for an incubator in Montreal (designers: D. Boisvert, A. Achdjian).

It should be pointed out that this project also exemplifies an original approach to an urban renewal problem. As was mentioned, the abandoned factory that was destined for this project lay alongside a recently upgraded linear urban park. The amenity this park represented was intended by the municipal planners to catalyze a series of urban renewal projects leading to an increased tax base (accompanied inevitably by the process of gentrification, through the construction of condominiums in the reno-

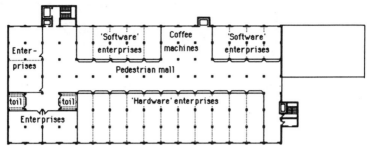

**3rd floor**

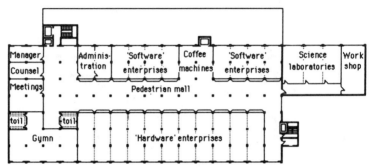

**2nd floor**

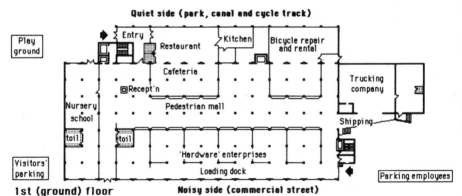

(b) Project for an incubator in Montreal (designer: Vital Chabot).

vated industrial buildings, for example). The local community feared this process, for the obvious reasons of increasing land values and consequent population displacement. Against this background, the incubator project was felt to be a way of upgrading the urban stock (thus fitting in with the character of the new park and also yielding higher tax income), yet maintaining convenient sources of employment in the neighborhood itself.

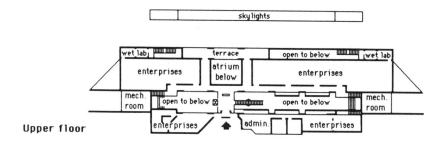

Upper floor

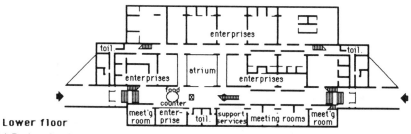

Lower floor

(c) Project for the Edmonton Advanced Technology Center (Barry Johns, architect).

The incubator shown in Figure 6-2c, however, was purposely designed to occupy a key location at the entrance to a technology park. The building is on two levels, set into the ground and 'turned into itself' in the sense that spaces for common activities are centered around the skylighted central circulation core and the atrium. The construction technology is simple, and materials are used in a way that reflects the laboratory-like character of the high-tech work carried out by the young enterprises. The project received a Canadian Architect's 1986 Award of Excellence. Areas break down as follows:

- Rental: 60%

- Services and shared spaces (including offices, meeting rooms, toilets, laboratories, food counter, etc.): 17%

- Circulation (including halls, stairs, elevators, etc.): 23%

The incubator shown in Figure 6-2d, the Chicago Technology Park Research Center, is a state-of-the-art wet laboratory facility. It provides 38 laboratories, seven at 400 sq ft (approximately 40 sq m), 10 with additional office and storage (100 additional sq ft, or approximately 10 sq m each), and 21 at 900 sq ft or approximately 90 sq m (also with additional office

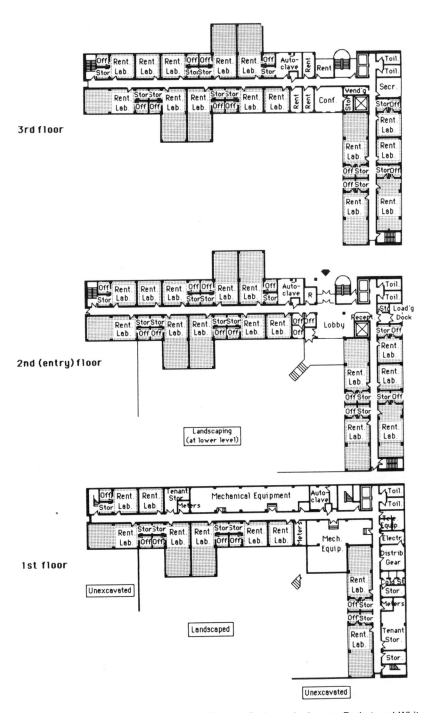

**3rd floor**

**2nd (entry) floor**

**1st floor**

(d) Chicago Technology Park Research Center (Graham, Anderson, Probst and White, architects).

116

and storage). High-tech features include: connections for fume hoods; independent exhaust systems for each laboratory; benches equipped with gas, air and vacuum lines; high-purity deionizing water supplies; darkroom; cold room; and autoclave dishwasher/dryer on each floor. Business and office support services include: office furniture in each laboratory, tenant storage areas (in addition to those adjoining most laboratories), conference room, word processing, photocopying, vending machines, and janitorial and security services. Tenants can also have access to research support services at the neighboring University of Illinois and to Rush–Presbyterian–St. Luke's Medical Center. Areas break down approximately as follows:

- Rental (including tenant storage): 58%

- Services and shared spaces (including offices, meeting rooms, toilets, vending and all mechanical and scientific support services): 24%

- Circulation (including halls, stairs, elevators, etc.): 18%

It must be pointed out that in these four examples, the proportion of rental to total nonrental areas (i.e., circulation plus shared services, etc.) is lower than in the majority of incubators. Indeed, in a few cases, a ratio of rental to nonrental of 90% to 10% is claimed by the incubator administrators (though, upon investigation, it is found that in these cases the shared spaces are counted as "rental spaces," and circulation areas are low because rental units are planned to occupy complete stories around a circulation core). The wide range of area ratios confirms once again the variety of functional programs to which designs must respond, reflecting in turn the importance of local circumstances in the programming process.

## CONCLUSION

The business and industrial incubator appears to be attempting the impossible: "providing more with less"—providing good-quality space and well-managed services at below-market costs. However, it can now be understood that these higher grade resources are *shared*, and it is the sharing that yields the economies that the fledgling enterprises so badly need. The policy planning and architectural design process must be systematically organized around this principle of sharing for cost reduction.

Once operating, an incubator conceived in this way offers to its tenants a new work environment, in which the individualism and competitiveness of the small enterprise are set in an environment of sharing and cooperation. Is this a paradox—or is it the precursor of a new work pattern?

## ACKNOWLEDGMENT

The author wishes to acknowledge the help of Elaine Gauthier and Dominique Oudot in the preparation of some of the background material for this chapter.

## NOTES

1. Y. Gasse, "L'incubateur: un outil de développement de l'entrepreneurship local," in *Les centres d'incubation d'entreprises*, Proceedings of a Colloquium, Université Laval, Faculté des Sciences de l'Administration, 1986.

2. H. Krist, "Innovation Centers as an Element of Strategies for Endogenous Regional Development," in *Science Parks and Innovation Centers: Their Economic and Social Impact, Proceedings of the Conference held in Berlin, 13–15 February, 1985*, ed. J.M. Gibb, (Amsterdam: Elsevier, 1985), pp. 178–188.

3. David L. Birch, *The Job Generation Process*, MIT Program on Neighborhood and Regional Change, (Cambridge, Mass: MIT, 1979), quoted for example by R. N. Cox, "Lessons of Thirty Years of Science Parks in the U.S.A.," in Gibb, *Science Parks*, pp. 17–24.

4. Illinois Department of Commerce and Community Affairs, and the U.S. Small Business Administration, Region V, *Starting a Small Business Incubator: a Handbook for Sponsors and Developers*, (Washington, D.C: Small Business Administration, Office of Private Sector Initiatives, 1984).

5. G. Haude, "The Role of Polytechnics in the Creation of Enterprises," in Gibb *Science Parks*, pp. 103–106.

6. C. H. Davidson, E. Gauthier, and D. Oudot, "Incubateurs d'entreprise—une stratégie de l'innovation," in *1st International Congress in France: Industrial Engineering and Management, a Key to Industrial Competition*, (Paris; Association française pour la cybernétique économique et technique, 1986), pp. 723-736.

7. J. M. King, G. F. Economos, and D. N. Allen, "Public and Private Approaches for Developing Small Business Incubators," in *Colloque 30ème conférence mondiale du Conseil International de la petite entreprise: la PME à l'ère entrepreneuriale, 16-19 juin 1985*, (Montréal; Conseil international de la petite entreprise, 1985), pp. 392-410.

8. P. Naylor, "High and Low Technology Businesses," in Gibb, *Science Parks*, pp. 246–249.

9. M. Temali and C. Campbell, *Business Incubator Profiles*, (Minneapolis; University of Minnesota, Hubert H. Humphrey Institute of Public Affairs, 1984), 130 pp.

10. J. Allesch, "Innovation Centers and Science Parks in the Federal Republic of Germany: Current Situation and Ingredients for Success," in Gibb, *Science Parks,* pp. 58–68.

11. David N. Allen and Janet Hendrickson-Smith, *Planning and Implementing Small Business Incubators and Enterprise Support Networks,* (Carlisle, Pa: National Business Incubation Association, 1985); see also: David N. Allen, *Small Business Incubators and Enterprise Development* (Carlisle, Pa: National Business Incubation Association, 1985).

PART THREE

# MANAGING AND DESIGNING PROJECTS FOR HIGH-TECHNOLOGY WORKPLACES

# 7

## CHAPTER SEVEN

# FACILITY MANAGEMENT STRATEGIES AND ORGANIZATIONAL CULTURE

## *FRANKLIN D. BECKER*

**H**igh-technology organizations are facing enormous challenges today. These come not only from intense national and global competition but from rapid changes in the facilities and equipment needed to remain at the cutting edge of technological and scientific discovery. Facility planners and managers, architects, scientists, and senior management are struggling to find the right combination of facilities design and policies for managing facilities effectively in a volatile business and scientific environment.

The problem is no less acute for university research laboratories. Competition for top-flight scientists, on which research universities' reputations rest, is fierce. To remain competitive and at the leading edge of scientific discovery, particularly during a period of shrinking financial resources, requires both appropriate physical facilities and equipment and

This material is based upon work supported by the National Science Foundation grant MSM-8405555. Any opinions, findings, and conclusions or recommendations expressed in this publication are those of the authors and do not necessarily reflect the views of the National Science Foundation. The author wishes to thank Professors William Sims and Robert Johnson, co-investigators for the study, for their contribution. Professor Sims was also a major contributor to the ORBIT-2 project.

innovative approaches to managing them. This chapter examines some of the issues involved in managing university research facilities and is based on a recent study of chemistry departments' research facilities at 40 major American colleges and universities. These findings are then related to a theoretical model linking particular types of organizations and organizational cultures to specific approaches to facility planning and management (1).

## UNIVERSITY RESEARCH FACILITIES

The capital costs of equipment and special facilities for sophisticated research is one of the most critical problems facing American universities today (2). In times of severe budgetary restrictions, not only are new construction and major renovation of research facilities curtailed but perhaps even more importantly deferred maintenance on existing facilities becomes the accepted way of coping with financial stress and avoiding cuts in payroll and operating budgets. The size of this problem alone is enormous. It has been estimated that in order to offset the cost of maintenance deferred nationwide since the 1960's as much as $35 billion may now be needed (3). Estimates of $4 billion for meeting current requirements for new equipment have recently been made (4). Between $15 and 20 billion may be needed to build and renovate university buildings and laboratories (5). In reviewing such statistics, Erich Bloch, director of the National Science Foundation, concluded that "the U.S. research system lacks adequate mechanisms and resources to maintain its infrastructure. Cumulative neglect has led to shortages of manpower, equipment, and facilities, in turn leading to policymaking and remedial action under crisis rather than to thoughtful planning for the future" (6).

The critical point of these statistics, and others that have been marshalled recently (7,8), is that our university research facilities are in danger of becoming obsolete. The fundamental issue, of course, is not facilities per se, but the extent to which they facilitate or impede the research enterprise. To continue to compete successfully with the private sector and with scientists around the world, American university research laboratories must not become a dead weight impeding scientific progress. It is argued that more federal dollars are needed, and efforts by the university community to secure more federal funds to upgrade research equipment and laboratories are beginning to get the attention of Congress. Representative Don Fuqua (D-Fla.), chairman of the House Committee on Science and Technology, for example, has introduced legislation that could pump an estimated $9.5 billion into U.S. college and university facilities between 1987 and 1996 (9). Many states are also beginning to respond with increased allocations, but is this enough?

## CAN FACILITIES MANAGEMENT HELP?

Is the issue simply more dollars, or is it also better use and management of existing space and facilities, as well as better planning of new facilities to make these more effective over a longer period? Stimulated by relentless competition from Japan and other international business competitors, American business has subjected itself to intense self-examination for the past few years in an effort to better understand what it does right and wrong and can change, to improve quality, contain costs, and remain competitive. The development of more sophisticated facility planning and management tools and techniques, intended to cut or contain costs while maintaining or raising building and organizational performance, has been an important facet of this general organizational reassessment. Colleges and universities need to begin asking the same questions. How can university research facilities be better planned and managed to meet the increasing pressures for more space and equipment within a tightly constrained economic context? What innovative policies and practices have been or might be applied to the management of university research facilities to improve space and equipment utilization, operations, maintenance, and ultimately research performance? Cornell University and the University of Michigan recently explored these kinds of questions in a joint study sponsored by the National Science Foundation.

## THE STUDY

The study focused on whether differences in research performance, as measured by such standard indicators as departmental rankings and number of publications and research funds per faculty, were related to scientists' evaluations of their research facilities and to variations in strategies and policies for planning and managing academic research facilities. Previous studies have focused on equipment and to some extent facilities (4,10), but to our knowledge no studies have focused on the relationship between university *facilities management policies and practices* and departmental research performance.

Forty major American colleges and universities participated in the study. Because the scope of university research facilities is enormously varied and far too complex to analyze in a single, relatively small study, we chose to concentrate on chemistry research facilities. Chemistry was selected because it is a capital-intensive research area, major changes have occurred in chemistry research methodology with strong implications for facilities (in particular, fewer wet laboratories and more experimental work with electronic equipment that has special facility requirements), and it is an established discipline occupying a wide range of new and older facilities. Preliminary site visits indicated that changes in methodologies and equipment requirements were generating new pressures on facilities, and we wanted to explore whether these pressures were stimulating new ways of planning and managing the research facilities.

The chemistry department rankings were taken from those developed by the 1982 report of the Conference Board of Associated Research Councils (11). The Conference Board ranking used a number of indicators, including number of publications and research dollars per faculty member, to avoid a single criterion. Our intent, in any case, was not to distinguish between any two or three adjacent rankings; rather, it was to distinguish two sets of research universities: one that by virtually any criteria would be considered outstanding and the other that would be considered strong, but not at the same level as the first.

Four sources of data were used in the study. A series of wide-ranging but focused telephone interviews were conducted with from two to four administrators, including the chemistry department chairperson, vice president of research, vice president of finance, and director of facilities at 30 (15 "outstanding" and 15 "strong") colleges and universities. The telephone interviews explored the way in which facilities planning and management was organized, key facility-related issues and concerns, and innovative policies being tried or considered to deal with research facility–related problems. These were followed by a brief mail survey to the chemistry chairperson and the university director of facilities, which was sent to 39 schools. The mail survey focused on several facility strategies and practices that we hypothesized might be related to research performance, and also asked about particular frustrations and innovations in facility planning and management. In addition to the administrator surveys, mail surveys were sent to faculty engaged in research to get a sense of the individual scientists' views of the facilities problem; 678 faculty members (55% return rate) in the chemistry departments at the 40 target schools responded to the survey. Finally, eight site visits were conducted in the chemistry departments at schools representing both the "outstanding" and "strong" chemistry departments to provide still more in-depth information including visual images and specific examples of university research facilities.

## PROBLEMS WITH UNIVERSITY RESEARCH FACILITIES

In general, the nature of facility-related frustrations identified by scientists and research and department administrators came as no surprise. No major differences emerged between the top- and lower-ranked universities in terms of the kinds of facility-related frustrations experienced. There were differences, however, in the extent to which higher- and lower-ranked departments experienced the same kinds of problems. Insufficient, inadequate, and inflexible space headed the list of problems. Environmental controls, in particular ventilation and heating and cooling of people and equipment, were major issues, as were high energy costs, building and equipment operations and maintenance, and electrical capacity and distribution to cope with the enormous influx of electronic equipment. A common theme was that there was often little awareness or understanding among federal and state officials that initial equipment and facility purchase costs represented the proverbial tip of the iceberg. One study estimated that the initial purchase cost of equipment represents about 25% of the total cost of operating a major piece of equipment, including staff time, maintenance, energy, and other material costs. Department heads, especially, were concerned about finding and funding qualified support staff to operate and maintain increasingly sophisticated equipment. They were also concerned about graduate students being pressed into the role of service technician to cope with equipment maintenance problems that increasingly involve sophisticated equipment requiring constant fine-tuning and specialized knowledge. Other issues raised by department heads were how to reduce space allotments, especially to older faculty whose productivity may have declined, or simply to make way for younger faculty the department wants to encourage and retain.

Directors of facilities at almost all the schools reported struggling with problems of insufficient and (especially in the lower-ranked departments) poor-quality space, and inadequate budgets for maintenance, repair, and renovation. Departments in state universities were particularly frustrated by the slow and cumbersome legislative bureaucracy that increased response and administrative time for securing approvals for new equipment and for renovation and new construction of facilities.

Many problems were related to characteristics of the facility planning and management process, particularly the relationships between the departments and central facility planning functions. For example, in several instances equipment would be ordered, at the departmental or individual scientist level, without its special utility and service needs, or its compatibility with existing or planned changes in building services or other equipment purchases being taken into consideration. Such piecemeal and uncoordinated purchasing decisions—typical in the highly autonomous culture of research scientists—cost time and money. In other cases building design and construction proceeded without there being a good understanding of equipment requirements, and also resulted in costly facility changes or mismatches between equipment, space, and services.

Among state universities, the excessive time needed to fund, design, and construct new facilities or make major renovations was a frequent complaint. As a result of long delays, user requirements often changed or new technologies became available that were not incorporated into new construction or major renovation. In some cases new buildings were outdated prior to occupancy. Administrators at state schools complained about state legislatures' failure to understand the need for new facilities, major renovations, and repairs. Shortage of funds and intense competition for them at the state level, combined with highly bureaucratized state procurement policies, contributed to the difficulty in making timely renovations and new construction. Administrators also reported that lack of long-range planning, including policies that would dedicate a percentage of overhead funds or operating expenditures to equipment and facility maintenance, repair, and replacement, further compounded these problems. Several respondents noted that raising funds for new construction, major renovations, and new equipment is more glamorous, and easier to achieve, than fund-raising for more mundane maintenance and repair needs.

## FACILITIES AND RESEARCH PERFORMANCE

As already noted, all the colleges and universities sampled experienced similar kinds of problems. The differences were ones of magnitude. Through the telephone interviews it became clear that facilities play a crucial role in departmental performance and that deteriorated or inadequate laboratories drive away good faculty and students while, as would be expected, high-quality facilities and equipment help attract productive research faculty and graduate students. One of our interests was in testing empirically whether the characteristics of facilities, including how they are planned and managed, relate to research performance. To examine this we related the individual scientists' assessments of their research facilities to the research performance of their departments, as measured by the ranking of schools as described above. Since it is the individual scientists who are most affected by the nature of the equipment and facilities support, we felt it made most sense to rate their personal assessments of facilities and equipment against department performance rankings and their own ratings of several research performance indicators.

The first task was to establish whether there were any differences between those universities ranked in the top 20 and those ranked 41 to 60. Figure 7-1 (a and b) indicates that the most significant differences between the two groups were related to support staff quality. Less important, but still statistically significant, were differences in the working conditions of the laboratories. According to the scientists' assessments, state-of-the-art equipment was more available and equipment and laboratory space were kept in better working order in the top-ranked than in the lower-ranked chemistry departments. Ratings of satisfaction with repair and maintenance of their laboratory and with the amount of workbench space were also higher in the outstanding than in the strong chemistry departments.

**(a)**

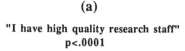

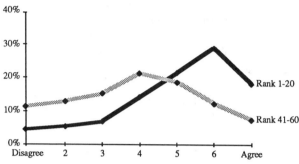

**(b)**

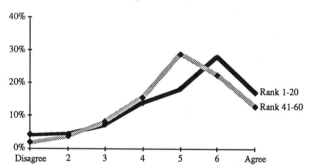

**Figure 7-1.** Relationship between departmental rank and quality of research staff and working condition of the laboratory.

The interpretation of these findings was consistent with what one would expect. Facilities play a support role in the university research establishment. Of primary importance are the number and quality of available faculty and staff. Instrumentation needs are also of primary importance, followed by the adequacy of the research facility. Thus, findings from the interviews, site visits, and survey suggested that the condition of research facilities can contribute to an understanding of the differences between the top 20 (outstanding) and the lower 20 (strong) departments.

The administrative interviews, in particular, indicated that an issue of central importance to the university research community was the relationship between the condition of the research facilities and the ability to attract and retain high-quality staff. One administrator reported that 25 faculty left the university in 1983, partly because of low salaries and partly because of poor research facilities. An analysis of the data from the principal investigator questionnaire was conducted to test if there was any evidence to suggest that poor-quality facilities were in any way associated with either the quantity or quality of research staff. The findings for the two groups are presented in the four graphs in Figure 7-2. The pattern of the relationships in all four graphs suggests that some association exists be-

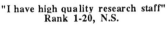

Rank 1-20

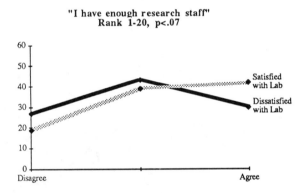

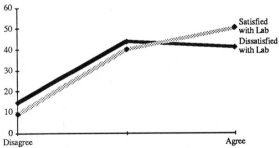

**Figure 7-2.** Relationship between departmental rank, satisfaction with laboratory and number and quality of research staff.

Rank 41-60

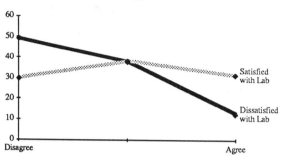

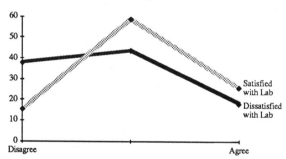

**Figure 7-2 continued.**

tween the amount and quality of staff and whether the principal investigators are satisfied with the condition of their laboratories. The relationship between ratings of quality of research staff and satisfaction with laboratory conditions was weakest in the top-ranked departments.

In summary, our data suggest that the condition of the research laboratory is associated with ability to conduct research and attract high-quality research faculty and staff. Top-ranked chemistry departments are more likely than strong departments to have higher quality staff, better instrumentation, and laboratories that are kept in good working condition. These findings were expected. Outstanding universities tend to have the ability to commit greater resources to research.

An important point is that the ratings of satisfaction with laboratory facilities and equipment for both outstanding and strong departments fall around a neutral point; that is, while scientists at top-ranked departments are more satisfied than those in lower-ranked ones, none of them are especially dissatisfied with what are, in effect, the tools of their trade. This is not to say that the amount and quality of space and equipment in top- and lower-ranked departments are the same—they are not. Site visits to chemistry departments revealed considerable differences in amount of space per researcher, number of researchers sharing major equipment, the general condition of the facilities, and the nature of support facilities ranging from departmental libraries and reading areas to seminar rooms, machine shops, and stock rooms. The top-ranked departments, for example, more often had extensive departmental libraries, stockrooms, and machine shops, and these were often accessible on a 24-hour basis. They are in several cases enormous operations. The departmental stockroom at one top-ranked department was a veritable supermarket, for example. It is a three quarter of a million dollar operation, operating in 5200 sq ft (483 sq m) of space, that stocks almost 3000 different items. One of the other top-ranked departments occupies 140,000 sq ft (13,006 sq m) of space, has 225 graduate students, 150 postdoctoral students and 33 faculty.

The difference in the quality of laboratory space is reflected in the survey data. Forty-eight percent of the individual scientists in outstanding chemistry departments reported that their laboratories had been renovated, compared with 26% in the strong departments. At one of the highest ranking of the outstanding departments we visited, only 2000 sq ft (186 sq m) of a total of 140,000 sq ft (1.4%) for the chemistry department as a whole, had not received a major renovation within the past 15 years.

Not all of the lower-ranked schools had poor facilities; one we visited had a new building and a considerable array of state-of- the-art equipment. However, the likelihood that major renovations or expansion would occur in these laboratories over time in the near future was not as great as it would be in top-ranked departments, especially for the stars of those departments around whom departmental reputations are formed and major scientific discoveries are often made. Faculty superstars are much more likely than other faculty to have state-of-the-art equipment and high-quality laboratory space since these are the basic conditions for their selecting and remaining at a particular university.

Taken together, the administrator and principal investigator surveys indicate that there are differences in the amount and condition of facilities and equipment at top-ranked and strong chemistry departments. Further, differences in satisfaction with facilities and equipment, and with the policies that affect their availability, condition, and use were related to a number of different performance indicators.

# FACILITIES MANAGEMENT STRATEGIES: CENTRALIZED VS. DECENTRALIZED APPROACHES

A second study area of this project was the investigation of the role that various university resource allocation policies might play in the maintenance of high-quality research facilities. During the course of the project, the research team began to conjecture that those facilities that were best maintained tended to belong to those institutions that had instituted some type of decentralized building management. Several models for decentralized planning have been proposed, with James Conant's "Every tub has its own bottom" probably being the classic application of this model. In a more contemporary version of this approach, the University of Pennsylvania argues that universities have grown too complex for effective centralized control, in part because central administrators tend to delay decisions in order to keep their options open as long as possible (12). In this study, the team was interested in understanding the effect of centralized and decentralized approaches to planning and managing facilities on research performance, as measured by indicators such as reputational rankings, number of publications per faculty, and others noted earlier. According to Hoenanck, the strength of a decentralized approach is that it places responsibility and control at a level where responsive planning can occur, providing operating units with positive incentives to husband their resources and spend them most effectively (13). We wanted to learn whether this occurred and whether it seemed to affect research performance in any way.

Initially, a major finding seemed to be the absence of a clear facilities philosophy guiding specific facility management policies and practices at all but one or two of the target schools. However, an analysis of the principal investigators' data showed that one of the factors that distinguishes the two groups is the degree of influence that faculty felt the department had in allocating resources for buildings and equipment. Higher-ranked chemistry departments were associated with a higher level of influence than were departments ranked lower (Figure 7-3). Chemistry research laboratories in which the researchers (either administratively or organizationally) feel they have some influence on the way resources are spent for buildings tended to be rated higher with fewer facility-related problems (Figure 7-4). A high level of influence was also associated with a tendency to be more satisfied with major laboratory renovation projects and departmental space allocation policies (Figures 7-5, 7-6).

Telephone interviews, and even more so site visits, suggested other major differences among institutions. At some universities the management of the research laboratories was thought of as a primary function of the central administration, while at others it was a primary function of the

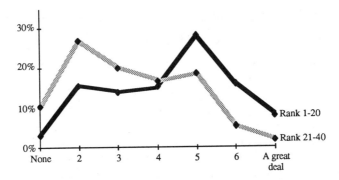

"How much influence does your department have in
allocating resources for bldgs and equip?"
$p < .0001$

**Figure 7-3.** Relationship between departmental rank and perceived departmental influence in allocating resources.

individual chemistry department. At universities with more decentralized approaches there were, in fact, strong principles guiding facilities decisions *at the departmental level.* These principles exist to varying extents in all university departmental facility decision processes, but they appeared to be honed to a fine level in the top-ranked departments, most of which were in private colleges and universities.

## Two Cultures

In state institutions, in which funds are appropriated by the state legislatures, the decision autonomy of the university in general, as well as of the

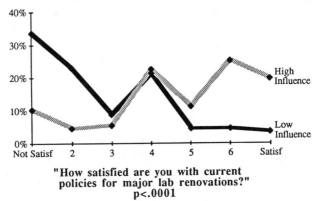

"How satisfied are you with current
policies for major lab renovations?"
$p < .0001$

**Figure 7-4.** Relationship between perceived departmental influence and overall quality satisfaction.

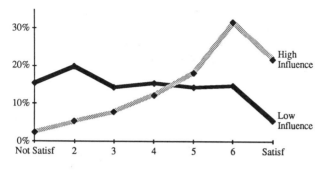

"How satisfied are you with the overall
quality of your current laboratory facility?"
p<.0001

**Figure 7-5.** Relationship between perceived departmental influence and satisfaction with laboratory renovations.

departments, is more constrained than in private colleges and universities. The opportunity for a department within a private institution to engage in its own facility fund-raising by issuing its own bonds, for example, is generally not available to a state institution. Rules governing salaries and wages and procedures for maintaining and allocating space also tend to be more stringent in state-supported institutions. These broad differences in administrative context are critical to keep in mind when looking at the freedom to make decisions at the departmental level in the top-ranked and strong departments.

At the top-ranked departments, the four key principles that guide facilities decisions such as space allocation and renovation, equipment acquisition, location, and maintenance *at the departmental level* reflect high levels

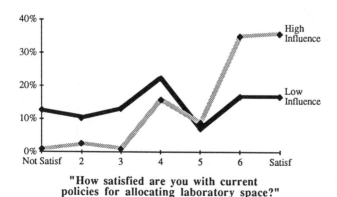

"How satisfied are you with current
policies for allocating laboratory space?"
p<.0001

**Figure 7-6.** Relationship between perceived departmental influence and satisfaction with laboratory space allocations.

of freedom to make decisions and organizational autonomy. The four guiding principles are: no explicit rules, benign dictatorship, productivity, and user responsiveness. In part, these principles are what enable the top-ranked departments to maintain their special position.

## No Explicit Rules

Explicit rules of any kind were a rarity (less than 15% on average) at the departmental level for all the top-ranked schools, whether for equipment purchase and maintenance, space allocation, or the purchase of special supplies. A major consequence is a high degree of flexibility. Exceptions are easy to make. New situations, whether a change in research directions or a decline or increase in the productivity of a research group, can be relatively easily accommodated because one does not have to pay attention to formalized rules and procedures that may, for example, require everyone at a given rank to have an equal amount of space. This policy is well illustrated by the treatment of "superstars" at the top-ranked departments.

SCIENCE SUPERSTARS

Top-ranked departments are in fierce competition for the same limited pool of "big guys." These science superstars, some in the midthirties, can and do make enormous demands for equipment, space, salary, and support staff, as well as reduced administrative and undergraduate teaching responsibilities. Their demands reflect and generate status, but they are fundamentally productivity driven. Office renovations, new furniture, and a seminar room are often part of the facility package offered to science superstars, but most important are more and new laboratory space and state-of-the-art equipment. In top-ranked departments, and particularly at the highest-ranked departments, the organizational culture (which reflects the basic shared values of the group's members) accepted a departmental commitment to inequality in space and facilities allocation among faculty. In a world of finite resources, the superstars took more than their share of all resources. In several top-ranked departments, this meant the most productive and prestigious scientists' having as much as a full floor (in one case almost 20,000 sq ft [1858 sq m]) of space, compared with others' within the same department having from 1200 to 5000 (111 to 464 sq m) sq ft. At the top-ranked departments these disparities are both possible and tolerated by the faculty because such colleagues' scientific productivity and reputation enhance the department's reputation, and hence their own. Major equipment used on a shared basis also attracts and can benefit faculty who otherwise would receive a smaller share of resources. It should be emphasized that the smaller laboratories within such departments are not substandard, but simply represent a smaller-scale research effort.

What seemed to set top-ranked departments apart from strong departments was their ability to meet the demands of these science superstars. In the superstar culture, many facility decisions, from laboratory design to space allocation, are superstar driven. The absence of explicit rules makes this kind of flexibility possible. Department heads, and in some cases university presidents, are able to effectively recruit and keep top scientists because they can make decisions relatively quickly concerning the allocation of resources. They are also able to make effective faculty resource and staffing decisions with a minimum of outside interference or adherence to stringent rules and can, if deemed necessary, by-pass more collegial and democratic decision making.

## Benign Dictatorship

The benign dictatorship principle recognizes that final decisions rest with the department chair, whose power rests ultimately on the principles of *productivity* and *user responsiveness* (see below); that is, the commitment to exercise power to push the research endeavor (and the departmental reputation) forward. Benign dictatorship requires departmental autonomy, strong consensus about organizational objectives, and clear performance indicators. Seventy-two percent of the department heads reported that space allocation decisions were made on an ad hoc basis, *often by themselves,* after informal consultations with faculty. This reflects the autonomy and power of department heads within their own administrative units in most academic contexts, and especially in top-ranked departments.

The reluctance to exercise power unilaterally was reflected in one major chemistry department where an "innovation" was the appointment by the department head of a space committee to assist him in space allocation, especially space reassignment among older faculty. Reducing laboratory space for a scientist was likened to castration in its emotional upheaval. The space committee, headed by an older faculty person who was selected because he used little space himself and was widely liked and admired, was an administrative device to share the responsibility for making unpleasant decisions.

The space committee as an innovation illustrates that the departmental head's power rests on a base of political support that requires that the members of the department agree not with every decision but with the overall pattern of decisions as they relate to agreed-upon organizational objectives. This position also often rotates among senior faculty. Under this system, one is careful not to do unto others what one does not want done to oneself at a later point in time. Final decisions rest with the departmental chair, but decisions at this level are a last instead of a first resort.

The chair exercises the formal power of the position only when informal negotiations fail. Interviews and site visits suggest a pattern in which a faculty member with a space request will approach the department head or the executive director charged with handling the everyday administrative responsibilities of the department. This scientist will be told to talk with another faculty member whose research program has failed to use some of the underutilized space in his or her laboratory. The problem in many cases is resolved at this level, because of the third and fourth principles, which are discussed below.

## Productivity

Productivity, above all, is valued and rewarded in outstanding chemistry departments. Status differences (in some cases a veritable caste system) are integral to top-ranked departments, and status is linked to productivity and the magnitude of the scientific contribution. Scientists are driven by their work. What they demand, above all else, are good equipment and facilities. They are independent and highly autonomous. They want to be consulted about major departmental policy issues, but they shun routine administrative responsibilities that detract from their research. They will accept environmental conditions and salary structures below those of comparable private sector research units, but they will not tolerate—especially if they have other options, as most first-rate scientists do—facilities and equipment that impede their quest for new discoveries. Academic reputations, and increasing fortunes, depend on making important breakthroughs and making them before competing laboratories down the road or across the ocean do. While no scientists want to give up any of their laboratory space, in the face of a more productive research group's obvious space needs, such agreements are often hammered out between scientists and between scientists and departmental administrators. These may take the form of short-term "space loans" with the clear understanding that if the research funding and productivity of a group increases, it will want the space back. *These* agreements (in contrast with most departmental and university policies) are often explicit and in writing. If such informal negotiating does not resolve the problem, the decision will fall to the department head. While the decision is unpleasant to make, if it is obviously made on the basis of a productivity criteria, it is likely to be accepted and respected.

Within this culture, with its unrelenting emphasis on performance, the superstars have enormous prestige and power. University as well as departmental reputations are built on such "names." Hence the enormous bargaining power of science superstars and their departmental administrative advocates. Lower-ranked departments have faculty who share the same attitudes about work just described, but because no one or two persons stand head and shoulders above the others, and because there is likely to be less departmental autonomy (especially in state schools), the willingness of colleagues to accept disproportionate allocation of scarce resource is less in lower-ranked departments than it is in the very highest ranked ones. The result, in lower-ranked departments, is a more egalitarian attitude.

## User Responsiveness

The faculty's acceptance of the departmental head's power rests, as noted above, on the fourth principle of top-ranked departments: user responsiveness. One aspect of this principle is the willingness to try to provide for the differing needs of individual researchers. Custom-designed laboratories are an example of this. Another aspect of this principle is the department head's commitment to "clear the decks" by anticipating most needs and providing for them, and by reducing extraneous administrative responsibilities and involvement so that scientists can devote their full energies to research. In the top-ranked departments with a strong superstar culture, user responsiveness entails doing whatever is necessary to attract and keep the potential Nobel laureate—from salary and secretary to laboratory, office, and seminar space. In the lower-ranked departments, space allocation is likely to be more egalitarian, and faculty are less likely to be excused from routine administrative responsibilities or from heavy (relative to those of superstars) teaching loads. Again, in part this is because of the more bureaucratic and rule-oriented funding and administrative context in which such departments often operate.

The ability of top-ranked departments to be flexible about meeting faculty needs was demonstrated in several examples provided by administrators during site visits. At one top-ranked department, for example, the director of the laboratories is considered a genius at equipment repair. He is so good that because of him the department does not need to maintain service contracts on equipment. His salary reflects his worth, and it is not constrained by an explicit salary schedule tied to the position level. Part of the savings from unnecessary service contracts goes into his salary, and part of it can be used for other purposes, including reducing the user rates for the equipment. Beyond financial considerations, less time is spent on equipment maintenance and repair, and this is a critical concern to scientists. His replacement is currently being trained so that his retirement will not upset the current situation.

This same department's executive director, a full-time administrator who reports to the department head and who is responsible for carrying out the day-to-day administrative functions of the department, provided another example of decisions that reflect a clear-the-decks approach. An existing departmental library, which also served as a reading room for students and which duplicated many of the holdings of the main science library next door, could not support the necessary full-time librarian. An agreement was made whereby in exchange for the chemistry department's giving up its librarian, the main science library would stay open on a 24-hr basis, using the librarian's salary to fund the longer hours. This kind of bargaining is not unusual. At a different top-ranked department the demand for additional space was met by bargaining with the local city officials for air rights over a street that could be used to connect existing buildings. In exchange the university built a pedestrian underpass that increased safety. In another department "float," or spillover, laboratory space is maintained in some of the isolated parts of the building. Faculty whose space needs exceed their regular laboratory are offered this less desirable space for expansion. It is also where junior faculty often find themselves.

In summary, these four principles work together to create a coherent facility strategy consistent with the organizational culture of the top-ranked universities. The strategy is flexible and responsive to user demands. It is well suited to coping with change, uncertainty, and limited resources. The key characteristics of the top-ranked departments, where these principles operate strongly, are the absence of explicit rules, unequal allocation of resources, quick response time, organizational autonomy, a strong chair, and current performance as a central basis for resource allocation decisions. At the lower-ranked departments, where the four principles operate less strongly, decision making tends to be more bureaucratic, slower, more democratic, and less performance driven.

## FACILITIES AT THE UNIVERSITY LEVEL

While at the departmental level, particularly at the top-ranked departments, there was a strong, albeit informal, underlying philosophy guiding facilities decisions, at the university level this was less true. The kinds of policies we looked for initially at the college and university level, such as dedicating specified percentages of overhead or other parts of the university budget to maintenance, repair, renovation, and timely equipment replacement—all of which should reduce downtime—were virtually nonexistent. In fact, formally developed innovative policies of almost any sort were in short supply. Bureaucratic procedures and ambiguous and convoluted reporting and reviewing relationships between campuswide facility units and the departments, not typical *within* departments, contributed to facilities problems and frustrations.

In contrast to the absence of explicit guidelines at the department level, such explicit guidelines were reported to exist by 65% of the directors of facilities responding to the administrative mail survey about guidelines for major and minor renovation, new construction, energy management, and equipment maintenance. The telephone interviews revealed that most of the guidelines concern routine administrative sequences for approvals and procedures for purchasing goods and services. There were very few explicit or informal but widely shared procedures or criteria to guide major facility decisions such as establishing renovation or new construction priorities, programming of major facilities, or assessing the suitability of proposed plans. There were procedures for committees such as campus planning committees to review plans but no explicit criteria for such committees to follow in making their assessments. As is often the case, the most well-defined and assiduously carried out procedures often concerned the most trivial areas of payback.

Although there was strong informal consensus at the departmental level concerning organizational objectives for facilities decisions, there was less consensus at the university level. This is not to say that concern for energy conservation, health and safety, and facility flexibility were not widely shared objectives. Over the past 10 years these considerations have increased enormously in importance. They were not, however, an integral part of every facility decision, nor were the outcomes closely monitored, as they were (often by faculty themselves) at the departmental level.

Sixty-five percent of the facilities directors reported using outside consultants, primarily architectural and engineering services for major projects. In two cases—both private universities, one in the "outstanding" and the other in the "strong" category—use of outside consultants had gone much further. The strong school's entire facility operation is handled on a contract basis by a private firm. There is no in-house, university-funded central facilities administrative unit. In the outstanding school the only functions done in-house concern maintenance. Virtually everything else is contracted out, from planning and design to landscaping services, with individual sections of the campus having tremendous individual autonomy in their selection of outside vendors. The impact of such policies is unclear, a point we will return to in a moment.

In the remaining college and universities, where most facilities functions with the exception of major design and construction projects were done in-house, an important issue was the organizational relation between different facilities functions, in particular operations and maintenance and planning, design, and construction. Only 23% of the facilities directors reported some kind of "integrated" facility organizational structure in which basic planning and architectural design decisions were fully coordinated with the operations and maintenance of the facility. Very few central facility units were involved in any way with departmental "business" plans during developmental stages. Given some of the complaints about the inadequacy of new and recently renovated research facilities, particularly in terms of the cost of operating and maintaining them, this is of real concern.

Several schools related war stories about heating, ventilation, and air-conditioning (HVAC) systems and other building systems that operations and maintenance people found inefficient and difficult to maintain after occupancy. They felt these situations could have been avoided if they had been meaningfully involved early in the planning and design of major renovation or new facilities. Such seemingly little problems tend to have relatively large negative cumulative effects. Difficult-to-maintain HVAC systems, for example, tend not to be cleaned as regularly as they should be. System disfunction then causes significantly longer downtime than there would have been with a better-designed system initially. Research scientists are loathe to interrupt their research. Even a few days can be immensely disruptive and frustrating. Thus maintenance is likely to be deferred, leading to systems whose performance progressively degrades over time until the disfunction is sufficiently great to trigger a major maintenance or repair effort.

## DIRECT AND INDIRECT FACILITY EFFECTS ON PERFORMANCE

Unfortunately, life cycle cost issues related to facility performance, which universities increasingly consider in facilities decisions, generally focus on design and operational management policies intended to save energy and other operating costs. The broader and more fundamental issues of how facility allocation decisions affect the achievement of basic organizational goals, in this case research performance, tend to be ignored. When the effect of facility decisions on performance has been examined, in the context of office design for example, the focus has been on how the design of the proximate work environment—usually the workstation—affects performance (14,15). The concern has been with how the microdesign of the office environment facilitates or makes more difficult the performance of daily tasks and activities of workers, although these themselves are rarely measured directly. Our data suggest that the facility acts *indirectly* on performance through its ability to influence job acceptance and retention (Figure 7-7).

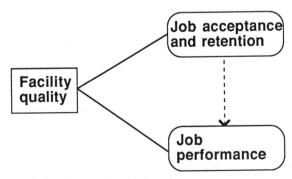

**Figure 7-7.** Impact of facility quality on job performance.

One reason the influence of facilities may be greater for molar behaviors such as job retention and acceptance than for daily activities stems from the culture of research scientists. They are an immensely dedicated, internally driven group of people who are devoted to their work. Typically, research scientists will work to overcome any obstacles within their power to conduct their research at a level they consider acceptable. Given this kind of motivation, the decision is not whether to work more or less assiduously (as it may be for many white-collar workers who are less internally driven and whose work is often far less intrinsically rewarding). The fundamental decision for research scientists revolves around taking a job and remaining in it. For top-flight graduate students, the decision is whether or not to enter a particular graduate program. Our data suggest that the nature of the research facilities can affect these basic work decisions.

In several cases research administrators talked about losing—or gaining—faculty and graduate students because of the amount and quality of laboratory facilities. It is the quality of these people that ultimately determines research performance and that helps establish departmental and university reputations. In the top-ranked and mostly private departments we studied this relationship is well understood. To make sense, life cycle cost analyses must take such subtle and indirect facility effects on organizational performance into account. Top-ranked universities do it implicitly already; they must to remain competitive. One of the problems in assessing the relation between facility costs and research performance is that the provision of research facilities has a minimum cost below which no organization can fall and still participate in the activity. There is probably a second threshold at which the increased cost associated with more and better facilities begins to be offset by increased productivity because of the better people attracted and the increased overhead on grants and contracts that they, in turn, attract (Figure 7-8). At what point above the minimum cost of facilities such added value occurs is an interesting empirical question.

The problem is made even more difficult by the fact that there is a time lag between the onset of facility degradation or improvement and performance degradation or improvement. Facilities degradation takes time. Its progress is slow and changes are often imperceptible. As a result, scientists do not immediately curtail their research or quit their jobs. While facility improvements are more likely to be noticeable, they may also suffer a time lag in their positive effects since many scientists are concerned with continuity of change—that is, how likely it is that a current renovation or addition will be followed, rather rapidly, by other improvements and continuous upgrading of facilities and equipment within succeeding years. One characteristic of top-ranked departments is that ongoing facility improvements are typical, and they generally are under the control of the de-

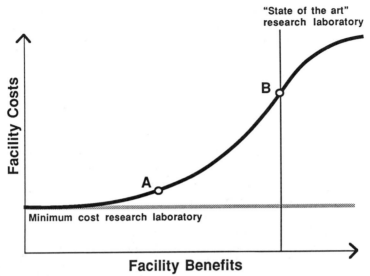

**Figure 7-8.** Laboratory cost/benefit curve.

partments themselves. The cost/benefit model in these institutions draws boundaries in assessing benefits that go beyond the cost-generating unit (i.e., the department) to include, implicitly, reputation as a major component of the benefit analysis despite the difficulty of causally linking facilities and reputation in a simple way. Such information may be difficult to quantify or to justify in public discussions, in contrast to information such as energy or maintenance costs, which are not only easily quantified but also more likely to be widely considered as legitimate factors guiding facility decisions. Nonetheless, such factors as reputation undoubtedly govern many facility decisions.

## INNOVATIONS AT THE UNIVERSITY LEVEL

Whatever the basis for actual facility decision making, the fact remains that university resources are limited. The challenge is to develop innovative policies for their allocation that maximize their value to the institution. In general, neither the administrative phone or mail survey nor the site visits identified very many campuswide innovative policies or procedures for addressing facility concerns ranging from deferred maintenance to renovation and new construction. Of the policies and practices in place at the university level that could be considered innovations, most dealt with problems of equipment acquisition and funding.

## Equipment Sharing

The most widespread practice was found to be the development of centralized major equipment centers both within the chemistry department (88%) and within the college and university (76%) for shared use of major equipment. Outstanding chemistry departments were more likely to have access to shared major equipment centers outside their department (70%) than were chemistry departments in the strong schools (47%). Only 36% of the schools had special equipment funds located within the department, compared with 50% of the schools that had such funds available collegewide or universitywide. Of those schools with a special college or university equipment fund, 70% of the top-ranked schools were represented, compared with 50% of the lower-ranked schools.

## Charge-Back Systems

Equipment charge-back policies, in which scientists use major equipment on a fee-for-service basis, were fairly common. In contrast, space charge-back systems, in which space beyond an initial minimum allotment must be paid for from research grants by the scientist using the space, were reported by only 26% of the facilities directors.

## Lobbyists

To improve funding 42% of the target schools (50% of the outstanding schools and 36% of the strong schools) had hired research lobbyists whose primary job function is representing university research to those outside the academic community in both the private and public sectors. Although a lobbyist per se says little about commitment to peer review, the position by definition is intended to influence potential sponsors, particularly in the private sector. While 60% of the schools contacted reported not having a special lobbyist, of those that did, a higher percentage were top-ranked schools.

## Separation of Research and Undergraduate Teaching

Several schools have physically separated graduate and faculty research in different buildings or have set up quasi-independent organized research units as a means of disengaging the research activity from teaching, particularly undergraduate teaching, so that academic scientists can devote more time to research. Administrators consider it easier to communicate the accomplishments and potential of such units to public and private sponsors, and they can provide an administrative mechanism that sometimes proves useful in dealing with state legislatures. Because such facilities are often new and emphasize research, they are also attractive to researchers, and can help in faculty recruitment. The right faculty, in turn, can affect funding levels significantly through its ability to attract public and private funding, including equipment grants.

## Other Innovations

Other innovations that the directors of facilities would like to see are general suggestions for dealing with the kinds of facility-related problems noted earlier, such as programs that would generate more maintenance and renovation revenue. There were several calls for "generic" laboratory space that would facilitate expansion and shifting of laboratory functions across disciplines and subareas within disciplines as program areas shrink and grow and change over time. This goes against the current trend of tailoring laboratories both for specific areas of study and often for particular scientists. Suggestions of specific policies, such as a research space "rental" system (space charge back) to encourage reassignment of unproductive space (and presumably unproductive scientists), were rarely mentioned. The main difference between the "outstanding" and "strong" colleges and universities was that the outstanding schools offered fewer suggestions for innovations. The kinds of suggestions that were made were essentially identical for all the schools.

The kinds of innovations suggested by department heads at the lower-ranked colleges and universities included increasing autonomy and discretionary decision making at the department level by returning a larger share of overhead to the department, including having a portion of research overhead directly available for discretionary use by the principal investigator for each grant; improving equipment inventories and maintenance through such devices as equipment banking (having all equipment listed and potentially available for use by researchers or for cannibalization for parts or as a variant having all unused equipment stored in an "attic" where researchers could actually see what is available); buying equipment on time; providing for maintenance of equipment in grants; and planning for equipment replacement over the long term. Here again the outstanding schools suggested fewer innovations, which suggests that they consider their policies and procedures for managing space adequate. The few suggestions for innovation that were made concerned centralized instrumentation centers, both at a regional and university level, and the need for federal grants to support the maintenance of shared equipment, based on the amount and type of equipment obtained in recent years. It was also suggested that funding agencies help with line item salaries for professionals needed to care for and operate instruments.

## WHAT CAN BE DONE?

The data suggest that at the departmental level clear, although informal, philosophies are guiding facility decisions such as space allocation, renovation, and equipment use. The top-ranked departments' autonomy is considerable, particularly in private colleges and universities. Central facilities groups appear to play a minor supporting role in these departments. In public universities and in the lower-ranked colleges and universities, the role of campuswide facilities groups appeared stronger. The absence of superstars clearly weakens the bargaining power of the individual academic units vis-à-vis central administration. In the lower-ranked colleges and universities, where centralized facilities units played a stronger role, there was little evidence of coordinated and innovative thinking about how to plan, design, and manage facilities to increase their performance.

In the face of such difficulties, one approach is to seek creative solutions to resource allocation so that existing and limited resources are used more efficiently and effectively. This has been done, to a certain extent, but most of the effort has focused on energy conservation and shared use of major equipment. Lobbying for increased federal funding and developing stronger ties with industry have also been high on the "innovations" list. Relatively little attention has been paid to the problem of space utilization of existing facilities or to ways of organizing facility functions and the processes for making major facility decisions to improve the quality of the decisions and of the space and facilities.

Minimal innovation in facility-related policies within institutions is predictable given the general state of facility planning and management in large organizations today. Stimulated by enormous changes in the amount and rate of takeup of information technology and changes in organizational patterns and worker expectations, facility planning and management is just emerging from a largely reactive, crisis-oriented, lower-level operations activity into one that is proactive, planning oriented, information intensive, and fully integrated with other organizational planning functions (16). Until quite recently, facility management was rarely connected to business planning, and changes in business plans were rarely examined for their facility implications and requirements.

In the private sector, facility planning and management is developing a professional identity and new approaches to planning and managing facilities that fuse concerns with buildings and building systems, new electronic technologies, and organizational practices and cultures (16,17). Part of this longer-term evolution is the development of increased accountability and a research tradition, including systematic design and policy evaluation. In few other fields with this much financial risk are decisions made essentially in an information vacuum. University research facility planning and management is no exception.

Few current facility policies and practices are systematically evaluated to test their effectiveness. The kind of close-range informal evaluation that occurs at the department level by full-time executive directors and their staffs happens less at the campuswide level simply because the persons involved in the decisions do not have to live on a daily basis with the consequences of their decisions (as is true at the departmental level). Decisions—from those about the organization of facility planning and management functions to those about processes for planning and designing facilities to encourage energy conservation—are rarely made as part of either a clear organizational philosophy guiding facilities decisions or a programmatic research effort designed to test different concepts and ideas for allocating facility-related resources including money, space, and personnel. It is inconceivable that professions such as engineering or medicine would proceed this way. The cost of making a mistake, in human lives, is simply too great. Such dire human criticalities rarely occur in facilities planning and management decisions, but scarce laboratory space in poorly used and new buildings that are energy intensive, provide poor air quality, and are inflexible to changing conditions have enormous financial implications. The magnitude of money involved comes into sharper focus when one realizes that the cost of start-up laboratory facilities for a *single* chemistry professor can range from $200,000 to $1 million. Making the right decisions when such sums are involved is not a trivial concern.

## AN EXPERIMENTAL APPROACH TO PLANNING, DESIGNING, AND MANAGING UNIVERSITY RESEARCH FACILITIES

What we need to do (in addition to lobbying for increased federal funds) is to apply the power of the scientific approach to identifying effective facility management strategies: in other words, to take an experimental attitude toward facilities planning and management. One could then begin to explore whether some of the innovations proposed by respondents in our study, as well as many others, might actually improve resource allocation decisions. What would be the impact of regional shared equipment facilities on the research activity? What additional funds for travel would it require to work well, and what amount of productive time would be lost for the scientist driving or flying between his own laboratory and the regional laboratory? Who would benefit, and in what ways, from returning a percentage of overhead directly to the principal investigator? Would more proposals be submitted? Would more money be allocated to equipment service, reducing downtime and improving instrumentation reliability and accuracy? Would a space charge-back system result in more efficient use of laboratory space? Would storage be handled more effectively, so that very expensive laboratory space would not end up being used in part as a storage room? Would such a system stimulate new design ideas concerning the location of graduate student and postdoctorate nonlaboratory office space? These are policies whose potential benefits make them worth trying, but they should be implemented for a given period of time, with clear goals, and then evaluated so that they can be preserved, modified, or abandoned depending on the results of the study.

## Characteristics of an Experimental Approach

The first characteristic of an experimental approach is what organizational psychologist Karl Weick has called the "small wins" approach to planning (18). Weick argues that trying to solve huge problems can be overwhelming and discouraging. The fruits of one's efforts are often difficult to discern. Identifying smaller parts of large problems and tackling them is likely to be more rewarding and successful because the ability to influence events at a smaller scale is greater.

Universities should continue every effort to influence the federal and state legislatures to secure more funds and to develop more flexible and dynamic processes for allocating funds so that the time lag in new construction and major renovation is reduced and more funds are allocated for maintenance and operation of facilities and equipment once in place. This is a long process. It not only should not prevent but in fact should stimulate equally active attempts to influence facility planning and management policies and practices at the campus level. This is where universities exercise the most control over their own resources, including budgets and personnel, and where the effort to initiate change is more likely to be successful, at least in the short run.

One way of applying the small wins concept to the planning and management of university research facilities would be to start, not with a new policy implemented campuswide, but rather with one initially implemented for a particular college, or even a particular department, on a temporary, experimental basis. The fact that the new approach is experimental and is being tried on a small scale can help reduce resistance. Participation should be largely voluntary. The director of facilities, the vice president of research, the vice president of finance, the deans, and the department heads might meet to identify common problems and brainstorm innovative policies and practices that might help solve the problem. Those colleges or departments that wanted to participate could then initiate discussions with their faculty about details of the program, how it would be implemented and when and how it would be evaluated.

Whatever outcome measures are selected should be generally agreed upon as valid by most of the participants in the study, as should the means of measuring them. These measures should be "culturally powerful" within the particular organization: that is, they may be relatively subtle but important indicators of enhanced productivity such as reduced downtime, lower repair costs, fewer outside repair visits, and fewer delays in equipment acquisition. Or they may be user satisfaction ratings or direct research measures such as number of papers published or grants received. Multiple measures should be collected so that, in the final analysis, an overall performance profile can be generated, rather than a profile that emphasizes only one or two factors such as initial or operating costs (19). Thus some of the outcomes may be financial, others may relate to quality of work or space and equipment, and still others may relate to scientists' feelings and perceptions about the facility and use patterns.

The likelihood that the outcome measures selected within a particular department or college will be similar but not identical to those selected by another college or department should be expected. However, because the experiment is intended to explore the feasibility, ultimately, of the innovative policy or practice campuswide, at least some of the measures should have face value validity for other members of the campus community who may review the outcome of the experiment. The nature of the outcomes one tries to influence will necessarily influence the length of the assessment period. If the new policy is intended to reduce maintenance costs and time, then 2 to 3 years may provide a sensible period for making comparisons with baseline data before the policy is implemented. If equipment acquisition or facility renovation issues are the focus, a longer period may make more sense because the incidence of these events is less frequent.

The power of these small studies could be further enhanced if similar policies were tested at several different colleges and universities simultaneously. The intent would not be to have one single, mammoth study (thus violating the small wins philosophy) but rather to be able to draw on and compare similar, though not identical, situations to see whether any patterns emerge across the colleges and universities. Differences in findings would help identify contextual factors that may influence the value of any given approach. Similarities in findings would increase confidence in the value of the approach. In either case the benefits would be real.

Done on a small scale, but broadly, an experimental approach to the planning, design, and management of university research facilities would begin to place the facilities administration departments at universities in a much stronger position for providing the highest level of support to scientific research within the academic community. The collaborative nature of identifying problems and developing solutions would ensure that the strong cultures of university and departments are not arbitrarily violated in the name of efficiency. What is the alternative? It is the likelihood that piecemeal, fragmented, and untested working assumptions and personal opinions will continue to drive major research facility decisions—a most unscientific way to run a scientific endeavor. The cost of mistakes is sufficiently high, and the cost of an experimental approach sufficiently low, to make an experimental approach an attractive strategy.

## FACILITY MANAGEMENT STRATEGIES AND ORGANIZATIONAL CULTURE

The discussion thus far has underlined the connection that exists between organizational cultures and facility planning and management strategies. By organizational culture I mean the often unstated but usually widely shared principles, norms, and expectations that guide both decision-making processes and work practices in organizations (20). Facility planning and management strategies are those formal and informal policies and practices used to plan, design, and manage the physical resources of an organization. These physical resources include not only the building and its mechanical systems, space, furniture, and equipment but also policies and practices governing how such physical resources are used, by whom, when, and in what ways. In essence, facility planning and management strategies are one way in which organizations communicate their organizational culture to employees, visitors, and the public.

The characteristic culture of university departments, with their emphasis on broad faculty participation in decision making, informal and unwritten policies, and the absence of formal rules, creates facility planning and management strategies that are appropriate to that culture but that may not be appropriate to other units within the organization. These departmental units may be at odds with highly bureaucratized campuswide facilities units, especially in the case of state institutions that are subject to high levels of legislative control. The challenge is to develop typologies of both different kinds of organizations (and subunits within them) and different approaches to facility management. The pioneering work of the ORBIT-2 team addressed the problem of developing organizational and facility management typologies.

### The ORBIT-2 Organizational Model

The fundamental premise of the ORBIT-2 project was that "one size does not fit all"—that is, that each organization requires information technology, buildings, and facility management strategies that fit its particular social and organizational contexts. One outcome of the project was the development of a model of different types of organizations (Figure 7-9).

The two dimensions by which organizations were differentiated were *nature of work* (routine vs. nonroutine) and *nature of change* (high vs. low). These two dimensions were selected because in combination they could be

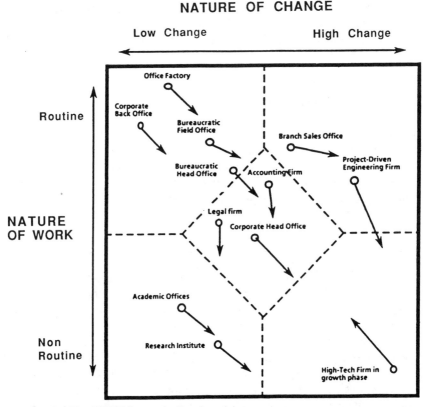

**Figure 7-9.** The ORBIT-2 organizational model.

used to describe departments (and in some cases whole organizations) at a level that was useful for thinking about facility decisions. Each dimension was intended to describe a general overall characteristic of an organization or subunit and it takes into account a number of specific aspects. For example, *nature of change* refers to the likelihood of internal reorganization, the frequency of relocating workplaces within the office, and changes in overall staff size. Each of these aspects of organizational change has obvious implications for design. For instance, to accommodate changes in overall staff numbers a building shell should be configured to permit expansion or subletting. To accommodate relocation of staff, interior fitting out should lend itself to rearrangement, rewiring, and rezoning of HVAC. Thus high-change organizations are likely to be quite different in their use of environmental resources from those that do not experience much change.

ORBIT-2 developed a series of survey instruments to measure these organizational characteristics, a system for rating building strategies for their ability to respond to some of the kinds of physical demands placed on buildings by different kinds of organizations, and a rating procedure for measuring organizational demands on facilities against the ability of the facilities to meet these demands.

The second dimension of the organizational model, *nature of work*, was defined as the extent to which the predominant work patterns within an organization or its subunits are either routine and predictable (e.g., claims processing, electronic filing, filling sales orders) or more varied and unpredictable (e.g., financial analysis, long-range planning, systems programming, market research, organizational development). These aspects, too, have facility implications. Organizations characterized by highly routinized work, for example, are much more likely to rely on a central mainframe and an omniscient management information system (MIS) department than is a research institute or high-tech firm in which computing intelligence is likely to be widely dispersed.

Combining these two dimensions describes a field of change in which the movement that all organizations experience can be traced (Figure 7-9). For example, work in the corporate back office, which is today highly routinized, may over time become less routine as new forms of job design and changing attitudes about computing emerge. The high-technology firm will probably trace an opposite trajectory as the firm matures, change slows down, and routine operations increase. One intent of the organizational classification scheme, with organizational change represented by the arrows, is to help organizations understand the direction in which they are likely to move over the next 5 to 10 years. More fundamental is the attempt to link these organizational characteristics to characteristics of facility management strategies over the coming decade.

## The ORBIT-2 Facility Management Model

Two dimensions were selected to characterize facility management strategies: The *amount of coordination among facility management functions* and the *degree of staff involvement* (Figure 7-10). Amount of coordination was selected because facility decisions increasingly require expertise from areas as diverse as mechanical engineering, architecture, planning, real estate, space planning and interior design, MIS, human factors, building maintenance and operations, human resource planning, and environmental psychology and organizational behavior. Maintaining effective facilities requires that such diverse knowledge be integrated into facilities decisions. The extent to which coordination among different departments and disciplines occurs, as well as the integration of expertise represented by outside consultants, becomes a major factor distinguishing facility management strategies.

Degree of staff involvement was the second dimension selected because the American work force is becoming more educated and more profes-

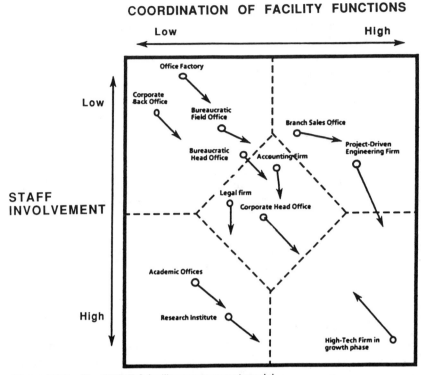

**Figure 7-10.** The ORBIT-2 facility management model.

sional. As it does, employees' expectations not only about the nature of work but also about the conditions under which work is performed are becoming more important (21). Several large research studies have shown that American workers want to be more involved in decisions concerning their immediate work environment than they did in the past (14,15). Such involvement has been shown to lead to decisions that better support workers' expectations and work style, increase commitment to and acceptance of design solutions, and result in more satisfaction with the workplace. Taken together, these two dimensions by no means exhaust the universe of factors that could be used to characterize facility management strategies, but they do capture what appear to be essential components of different approaches.

## LINKING THE ORGANIZATIONAL AND FACILITY MANAGEMENT MODELS

Used in conjunction with the organizational model described above, the facility management model just outlined suggests that for given types of organizations, some facility management policies and practices are more likely than others to result in facilities that meet the organizational demands placed on them (Figure 7-11). The ORBIT-2 project identified a number of specific components of these strategies. They include the organizational structure of facilities units and their involvement in corporate planning; the nature and scope of the facility group's decision authority; the nature of design programming and planning processes, resource allocation, staff involvement, environmental evaluation, and furniture and space policy; and the mechanisms for integrating diverse expertise into facility decisions. For each of these categories, a number of different options or approaches were identified and described.

Thus, based on the ORBIT-2 model, the nature of facilities strategies for an organization unit such as a university research laboratory, which could be characterized as nonroutine and relatively low change (compared to a high-technology firm in its start-up phase, for example), would have a high degree of staff involvement (at the departmental level) but relatively little overall coordination of facilities decisions between the department and campuswide units. This describes, in fact, the strategies (planned or de facto) at many of the university laboratories we studied.

## UNRESOLVED QUESTIONS

One interesting and unresolved question is the direction in which university laboratories are moving, in terms of the ORBIT-2 typology, and whether existing facilities strategies will or should shift over the next few

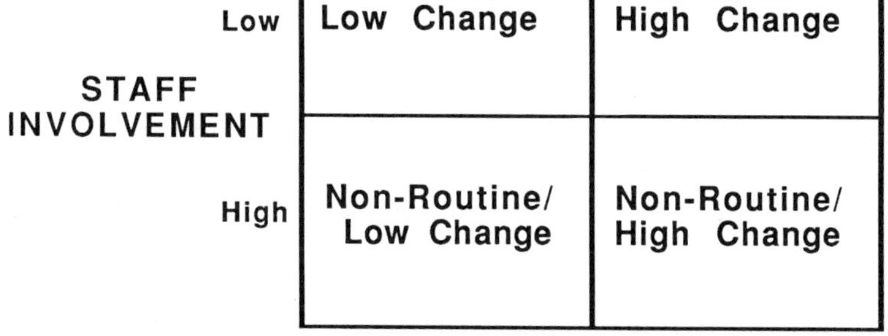

### COORDINATION OF FACILITY MANAGEMENT FUNCTIONS

| | Low | High |
|---|---|---|
| **Low** | Routine/ Low Change | Routine/ High Change |
| **High** | Non-Routine/ Low Change | Non-Routine/ High Change |

STAFF INVOLVEMENT

Figure 7-11. Linking the ORBIT-2 organizational and facility management models.

years in response to such changes. A second and critical question from the viewpoint of the ORBIT-2 project and the management of high-technology facilities in general is whether the implicit organizational and facility disfunctions predicted by the ORBIT-2 model will in fact occur. Do research organizations with facility policies and practices that restrict staff involvement and that are characterized by low to moderate degrees of coordination among different facility management functions actually perform more poorly? Are their spaces less adaptable? Are changes more expensive, disruptive, and time-consuming to make? The ORBIT-2 project made a significant stride in developing a model that linked organizational type to facility management strategy. The next step is to further refine and elaborate the model on the basis of empirical tests of its predictions.

A third unresolved question is the relation of different organizational subunits, each with different organizational styles and cultures, to other units within the same organization. The experimental approach, outlined above, is needed to systematically explore innovations in facilities organizational structure, policies, and procedures that can effectively bridge these different cultures and organizational boundaries. Can, for example, the development of strong guidelines by a campuswide or corporate group, within which various subunits can freely make choices without detailed review by the central body, meet both units of the organization's need for control and flexibility? Is a strong, well-understood, and clearly articulated corporate facilities philosophy more effective than written guidelines and rules? A recent study of facility management in large multinational corporations found, for instance, that such a clear organizational philosophy resulted in buildings built in different parts of the world having very similar characteristics despite the absence of detailed architectural guidelines and specification (22).

In conclusion, facility management procedures, practices, and organizational structures must fit their organizational context. The empirical challenge is to identify systematically which types of facility planning and management approaches work best for which types of organizations. During periods of prosperity, the costs of poor facility decisions are easily absorbed (and masked) because there is capital available to make expensive renovations, to lease or purchase more space, and to replace inadequate equipment and building systems. During periods of financial stringency, when the costs of mistakes are harder to absorb, making facility decisions that enhance overall organizational performance can be absolutely critical to organizational survival. As organizations grapple with new management styles and organizational structures to help them compete successfully, they need to devote as much effort to developing approaches to facility planning and management that fit the new organization as to the creation of new organizational structures and management styles.

# REFERENCES CITED

1. Davis, G., F. Becker, F. Duffy, and W. Sims. *ORBIT-2: Organizations, Buildings, and Information Technology.* New Canaan, Conn.: The Harbinger Group, 1985.

2. Kennedy, D. "Government Policies and the Cost of Doing Research." *Science* 227 (4686):480–484, 1985.

3. Kaiser, H. "Deferred Maintenance." In *New Directions for Higher Education: Managing Facilities More Effectively,* edited by H. Kaiser. San Francisco: Jossey-Bass, no. 30, 1980.

4. *Financing and Managing University Research Equipment.* Washington, D.C.: Association of American Universities, 1985.

5. Conference on Academic Research Facilities. Sponsored by the National Academy of Sciences' Government, University, and Industry Research Roundtable; the National Science Board; and the White House Office of Science and Technology Policy. Washington, D.C., Nov. 22–23, 1984.

6. Bloch, E. "Managing for Challenging Times: A National Research Strategy." *Issues in Science and Technology* (Winter): 20–29, 1986.

7. National Science Foundation. *National Academic Research Equipment in Selected / Science / Engineering Fields, 1982–83.* Washington, D.C.: National Science Foundation, 1985.

8. National Science Foundation (NSF 77–17). "Academic Science and Engineering: Physical Infrastructure." In *The NSF Science Development Programs: A Documentary Report,* vol. 1, chap. 2, 1977.

9. Crawford, M. "A $9.5 Billion Plan for Facilities." *Science* 229 (4708): 31, 1985.

10. Pelz, D. C., and F. M. Andrews. *Scientists in Organizations: Productive Climates for R & D.* New York: John Wiley and Sons, 1966.

11. Jones, Lyle V., Gardner Linzey, and Porter E. Goggeshall, eds. *An Assessment of Research Doctorate Programs in the United States: Mathematics and Physical Sciences.* Washington, D.C.: National Academy Press, 1982.

12. Morgan, A. W. "The New Strategies: Roots, Context, and Overview." In *Responding to New Realities and Funding: New Directions for Institutional Research,* edited by L. Leslie, no. 43. San Francisco: Jossey-Bass, June 1984.

13. Brill, M. *Using Office Design to Increase Productivity.* Buffalo, N.Y.: Workplace Design and Productivity, Inc., 1984.

14. Steelcase Inc. *The Steelcase National Study of Office Environments, No. II: Comfort and Productivity in the Office of the 80's.* Grand Rapids, Mich.: Steelcase, Inc., 1981.

15. Becker, F. and W. Sims. "Facility Management." In *Planning Office Space,* edited by F. Duffy, C. Cave, and J. Worthington. London: Architectural Press, 1976.
16. Gabler, E. "Is the Building 'smart' Enough?" *Building Economics* 1(1): 50–53, 1986.
17. Thomas, M. "ORBIT-2: How to Rate Facilities for Automation Phase-in." *Facilities Design and Management,* pp. 103–107, 1985.
18. Weick, K. "Small Wins: Redefining the Scale of Social Problems." *American Psychologist* 39(1):40–49, 1984.
19. Becker, F. D. "Work in its Physical Context: The Politics of Space and Time." *Proceedings from the Architectural Research Centers Consortium Workshop on The Impact of the Work Environment on Productivity,* funded by National Science Foundation. Washington, D.C., April 17–19, 1985.
20. Shein, E. *Organizational Culture and Leadership.* San Francisco, Calif.: Jossey-Bass, 1985.
21. Naisbitt, J. *Megatrends.* New York: Warner Books, 1982.
22. Becker, F., and J. Spitznagel. "Managing Multinational Facilities." Paper presented at International Facility Management Conference. Chicago, Ill. October, 1986.

# 8

## CHAPTER EIGHT

# MANAGING PEOPLE, PROCESS, AND PLACE IN HIGH-TECHNOLOGY INDUSTRIES

## *CECIL L. WILLIAMS*

In this chapter I will be discussing the managed facility. Previous chapters have focused upon particular aspects of the physical place or circumstances in which those places were developed. One thing that always strikes me in debates about facilities is how much attention goes into the aspects of building a building, site selection, building programming, and technological considerations. How very little time is devoted to what happens once they are built! It is actually the interaction between the employees, the work that they do, and the interaction with the facility itself that produces the outcomes necessary for successful corporate activity. The Facility Management Institute in Ann Arbor, Michigan, of which I was director until it ceased to exist recently (having accomplished the mission for which it was set up initially), used to center its attention upon that interaction. The term *facility management* was coined to embody the recognition that the physical place is a flexible tool that indeed can be controlled, changed, and utilized by the corporation that possesses it. In other words, the physical workplace becomes a true instrument of management. It is important that we understand the difference between facility management and what is commonly known in our business as *building management.*

Most people think of building management as those activities that address the mechanical and technologically oriented aspects of a building. It usually is directed toward keeping the building clean, warm, and cool and keeping the elevators and toilets working, all within prescribed budgetary constraints. There is considerably more to making a building "work" as a tool than most of us realize. Coordination among the many aspects is paramount. Decisions are required constantly. Management of these enormous assets is a must. I will present a scheme by means of which one can begin to think about some of these management principles. Space constraints prevent minute details because literally volumes could be written about the various principles of management from which one can approach the more specific *facility management.* One of the frustrations in setting up education programs of any kind is the consideration that so much must be grouped under the phrase *management,* but I would like to present one of those perspectives.

## HIGH-TECHNOLOGY INDUSTRIES IN CONTEXT

What is the context within which to consider high-technology industries? It has been established in the preceding chapters that high tech is not an entity in itself. It is not just having sophisticated machines. It is not simply accomplishing work through robots and lasers. It touches every phase of modern American industry because of new materials being used in products and processes, thus affecting costs of buildings and equipment, which in and of themselves may not look high tech. It is simply the next development in the building industry. High-tech industries are usually defined as industries that are producing highly technical equipment that will be sold to consumers. High-tech industries might also be seen as those utilizing high-tech equipment to manufacture relatively conventional products. In other words, our conclusion is that very shortly almost all work areas will, in fact, have been affected in some way by high technology, so the focus in this book might well be upon all workplaces, because the same general principles apply. Ultimately, all of us will have to deal with high technology because practically every industrial building will in some way have to deal with high-technology concepts.

To corroborate my thinking about the differences between high-technology buildings and others, I inquired into a data base developed at the Facility Management Institute a few years ago. This data base consists of responses from several hundred members of an organization called the International Facility Management Association (IFMA). The data base has responses from facility managers representing corporations located in the United States and Canada. The corporations range from very small organizations to mammoth organizations that report managing millions of square feet of space. From the data base, we compared people in high-tech industries (basically those in the electronics industry) with those in non-high-tech companies in order to find differences. Basically, we found very little difference between the two groups in terms of how buildings are actually managed. Frankly, in both instances, we found serious deficiencies in managing the people, process, and place interaction. There were virtually no differences between the high-tech and non-high-tech companies in terms of the manner in which they were staffed to accomplish facility management, very few differences in the ratio of staff to the number of square feet managed, and very little difference in the managerial reporting relationships. This supported our contention that high-tech industries are not a unity, and do not in fact distinguish themselves from buildings in other industries in any significant way. The fact that there are not differences, however, does accentuate some of the problems of facility management in high-tech corporations. It suggests that those industries are not insisting that their buildings keep pace with the technological changes in manufacturing processes and products. The high-tech industries continue to perform the new work processes in old buildings. Many buildings thought to be up-to-date when designed are obsolete by the time they are put into use because of the rapidly changing technologies that will go into them. Trying to adapt those buildings to accommodate the new technologies becomes a very expensive process. In other words, the building has not been seen as an integral part of the business plan of the corporation that inevitably includes whatever (in the way of equipment) is needed to increase corporate profitability.

Most discussions of high-technology businesses emphasize the product and the number of those products that can be made. In other words, the focus is upon the end result rather than the process that occurs in manufacturing a product. Attention is given to what that product itself will be able to accomplish in terms of speed of computation, number of man-hours it will be able to save, and other such considerations. The attention is not directed toward how a product will be made or how the building will be seen as a tool for the accomplishment of making a product. Little or no attention is given to the characteristics of the people necessary to carry out the process for the completion of the product. It is this amalgamation of attention—to the process, to the physical place, and to the people—that Facility Management Institute has called *facility management*. Usually the work process is the only activity that gets any sort of attention, and as mentioned previously even that activity takes second place in many high-tech corporations because of the focus on product capabilities.

## MANAGEABLE WORK SPACES

We are all familiar with the stories about start-up phases of high-technology inventions. Many of them are stories of people working in garages or in family basements, assembling pieces that ultimately result in a major breakthrough in electronic equipment. The inventor pays little attention to the physical environment in which that work is done. But the cumbersome environment in which one works at home or in a garage is superseded by the fierce, competitive desire to produce something revolutionary. Little attention is needed to make the environment support the activity. The work process and the characteristics of the worker override the poor quality of the physical workplace. Many of these endeavors have been successful beyond belief. Trouble begins to develop, however, when that initial one-person shop begins to grow and there is a necessity to produce multiunits. Ultimately, the entire process will be transferred to larger corporate headquarters with fine facilities, shiny laboratories, and people to do the job who may not have any investment in performing well. The new facility will have taken on characteristics pictured in magazines or copied after other known facilities that are probably not directly related to the initial thrust of the inventor. Productivity tends to go down in these instances. Innovation especially suffers. The physical place has become disconnected from the process of developing a new product and the fervor of the engineer making that product. Attention to the physical place and the work force becomes secondary, and as a result, productivity tends to decline and the quality of the product is likely to suffer as well.

The point I want to make in this example is that the building has not been seen as an integral part of the business plan of the new corporation. If we are going to provide adequate facilities for high-tech industries, work spaces must become as much a part of the goal as any of the other managed assets. The building must be seen as just as vital an asset—a tool if you will— as any of the materials, money, or personnel. When buildings are not seen in that light, they tend to be viewed only as costs and are not seen as contributing significantly to the productivity goals of the company.

One ray of hope in all of this confusion is that facility managers are becoming more aware of the need to manage buildings in ways that accommodate all of the factors that contribute toward productivity. Recently I saw in an airline magazine that architects themselves are beginning to talk to clients about "user-friendly" buildings. (Computer jargon is invading even the architecture community!) To summarize, our goal should be to work for the accomplishment of productivity goals in the same way any other piece of equipment will work to help the company make money.

What are the elements of manageable work spaces? The four factors that typically go into the creation of a "user-friendly" environment are:

1. Technology and computers

2. Facilities and furniture

3. Work and job functions

4. Social and people concerns

## MANAGEMENT STYLES, CORPORATE CULTURE, AND THE WORKPLACE ENVIRONMENT

In most high-tech industries, the concentration is on technology and computers only. The major ingredient that is lacking in trying to develop a model for understanding the interactions between all these activities is the context in which these activities occur. In other words, we need to come to some understanding about the culture in which technology, people, and work process will occur. A new appreciation for the necessity of understanding corporate culture has been developing for the past few years, perhaps actually culminating in a book published in 1983 entitled *Corporate Cultures,* by Kennedy and Deal. For our discussion, what we mean by *corporate culture* is simply a way of thinking about how an organization goes about doing its business. A more formal definition is that it is the formal and informal patterns of behavior—the beliefs, values, philosophy, history, and myths—that explain how "things are done around here."

All too often the assumption is made that all corporations are basically alike except that they manufacture different products or provide different services. I personally think that this attitude stems from the great American belief that everyone is "average" and that people themselves are more alike than different. Architects have made an attempt to remedy this sameness by designing the outside skin of buildings to differentiate corporations from one another. When one walks through the front doors, however, one becomes aware immediately that things inside the building are almost exactly the same as in the building across the street. Only recently have we begun to study organizational differences and how those differences should be reflected in building interiors and their management. The previous chapter, by Franklin Becker, is one example of such work; this chapter is another.

In addition to the recent work of the Facility Management Institute, similar ways of approaching management styles and corporate cultures have been developed by others. I am especially indebted to the work of John R. Kimberly and Robert E. Quinn and their book *Managing Organi-*

*zational Transitions.* Prof. Ian Mitroff at the University of Southern California and Dr. Warren Keegan, a private management consultant, wrote of similar models. All of these approaches are based upon the seminal work of Carl G. Jung, a Swiss psychiatrist who wrote during the early decades of this century. The model is presented in Figure 8-1. The horizontal axis of the model expresses whether the focus of the corporation tends to be within itself in a very "today" orientation with little attention paid to outside competition or toward a "broad sweep" of the outside world, being very aware of the larger competitive market and perpetually looking toward the future. This dimension offers the opportunity to place companies along a continuum from what is called the "internal" approach (within the company) to the "external" approach (outside the company). The vertical axis is made up of the two extremes: "control" and "flexibility." The control end of the vertical dimension describes those companies that value order and linearity, and that use rational decision-making processes leading toward organizational stability and consistency. The opposite end of that continuum is the company that values individuals and favors participative management in order to respond quickly to new opportunities. When the characteristics of corporations are plotted on these two axes, four different clusters of corporate cultures result.

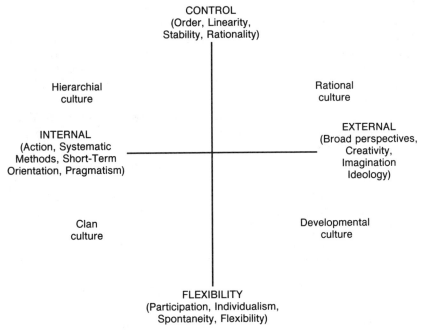

**Figure 8-1.** Model for corporate culture based upon control/flexibility management orientation and internal/external focus of management actions.

The culture described in the upper-left-hand quadrant is what Kimberly and Quinn called the *hierarchical culture*. It is characterized by

*assumptions of stability. Here it is believed that people will behave appropriately or comply with organizational needs when roles are formally stated and reinforced by rules and regulations. Surveillance and enforcement are important. In this legalistic view . . . the individual's primary need is for security and order. Leaders . . . [see] that the basic structure is maintained. As decision makers, they should proceed slowly, objectively documenting the process, thus ensuring the accountability of the outcome. It is assumed that the effective organization emphasizes stability and control and that the ideal form is the hierarchy, pyramid or bureaucracy.*

The quadrant on the upper right that is labeled the *rational culture* is

*permeated with assumptions about achievement. It is believed that people will comply with organizational needs if individual objectives are clarified and rewards are predicated on accomplishment. Compliance thus flows from formal contracting such as the MBO process. . . . The individual's primary need is to demonstrate competence and experience the successful achievement of predetermined ends. Leaders should be directive and goal-oriented, constantly providing structure and encouraging productivity. As decision makers they should proceed with reasonable haste, making logical, efficient decisions which are conclusive. . . . There is no place for wavering. It is assumed that the effective organization places emphasis on planning, productivity, and efficiency, and that the ideal structure is the market form or "theory A" organization of [Ouchi].*

The quadrant in the lower left of the model, which combines an internal focus of management with a flexible orientation, is identified as the *clan culture*. It is permeated by

*assumptions about human affiliation. It is believed that compliance flows from trust, tradition, and long-term commitment to membership in the system. . . . The individual's primary need is for attachment, affiliation, or membership. Leaders should be participative and supportive, showing consideration and facilitating interaction. As decision makers, they should proceed slowly, ensuring that all affected people participate and that the solution is supported by all members of the system. . . . The effective organization places an emphasis on the development of human resources and . . . the ideal form is the clan, family, or [Ouchi's] "theory Z" organization.*

The final quadrant, in the lower-right-hand-side of the model, describes the *developmental culture,* which is

> *permeated with assumptions of change. It is believed that people will comply with organizational needs because of the importance or ideological appeal of the task. . . . In this dynamic view, it is believed that the individual's primary need is for growth, stimulation, and variety. Leaders should be inventive and risk-taking, paying particular attention to envisioning new possibilities and acquiring additional resources, visibility, legitimacy, and external support. As decision makers, they should proceed rapidly, making the most legitimate decision, later adapting the conclusion as more information is collected or as external conditions change. . . . Compromise is important. It is assumed that the effective organization emphasizes growth and resource acquisition and that the ideal form is the adhocracy, organic system or matrix.*

Clearly, the four kinds of organizations created in this model differ significantly from each other. These four "pure" cultures do not describe every organization exactly. These patterns of organizational culture, however, provide a means for understanding the complexity of high-technology organizations and how their differences require unique responses from the facilities occupied by those organizations. Since our concern is the design of high-technology workplaces, the designer, architect, planner, or facility programmer must first determine the kind of culture in which he or she will be working. The culture of that organization will, in large part, determine the management of how people work, what kinds of people are hired, and what kinds of buildings are built. And as I have indicated throughout, the most important element in providing high-technology environments is whether the environments can be managed to accomplish the goals of the corporation.

This approach helps us move away from the unitary, monolithic notion that all high-tech buildings are the same. It helps the designer to move away from making assumptions about organizations based upon product characteristics. It provides a framework in which the environmental planner can utilize high-tech capability as a significant part of the corporate culture. In terms of the programming activities that must be done by architects and designers, this model provides an orientation to both the high and low technologies that must be utilized and planned for in order to satisfy the goals of the client.

The process that we are advocating in the design of high-technology environments is to begin by conceptualizing the project as the creation of a dynamic, manageable tool to be used by an equally dynamic work force. First, then, is the development of a working relationship with the manager who will be the primary liaison between the corporation and the architect/designer. It becomes the responsibility of the liaison person to be knowledgeable about the organization that he or she represents, what that organization wants to accomplish, what its rules and regulations are, and how to help gather information needed by the architect/designer. Both should have as their goal the providing of environments that enhance the entire constellation of activities that will occur there. If the building does not take on the characteristics of the business itself, that building is not going to be able to be the effective tool that the client expects it will be.

The notion that buildings should serve the *management* needs of corporations has not been widely promoted. The recent threat, however, from foreign markets (including the energy crisis) has brought a new emphasis upon profitability. Consequently, most corporations are willing to examine any aspect of the business that will improve productivity, and ultimately, profitability. As companies have explored these avenues, they have become aware that, at best, facilities have been overlooked. At worst, they have been unplanned, misused, and neglected. This is astonishing because buildings are the most tangible, concrete asset in the corporation. They have already been bought and paid for. Corporations began to realize that they were considering investments that had already been made but that needed to be leveraged in such a way as to contribute more significantly to the operation of the corporation. They needed to receive higher returns for previously made investments. It has been estimated, for instance, that in the United States 69% of the nation's wealth (in excess of $5 trillion) is invested in physical places that shelter and support human activity. It is not unusual for facilities to account for more than 25% of an organization's total assets. This is probably an underestimate for high-technology organizations because of the enormous amount of equipment that must go into those buildings. Yet in spite of the magnitude of this resource, corporations seldom think through the long-term organizational results they expect from facilities. Understandably, expecting little, they typically get just that. Because of the ever-increasing changes that go on in high-technology environments, facilities must become a catalyst for such change rather than an impediment to it. Management must begin to learn to use facilities as a way of changing how people work and of allowing work to be done more efficiently and effectively.

## THE FIT BETWEEN PEOPLE, WORK PROCESSES, AND WORKPLACE

Activities that go on within a certain corporate culture should be able to affect each other positively in order to contribute toward a productive environment. It is our contention that when there is a "fit" between the work process that must be done and the characteristics of the people who are hired to do that work, and when those two elements are placed in a negotiable, facilitative building built to be consistent with the goals of that organization, a high degree of productivity will result. Such a "fit" will result in higher job satisfaction, higher occupational satisfaction for the workers, higher productivity and profitability of the corporation, higher job performance, and greater satisfaction with the company.

The Facility Management Institute invested considerable effort to measure that fit between the three major elements of corporations before it recently ceased to exist, its objectives having largely been adopted by other organizations, such as the International Facility Management Association. There are many pitfalls in such an effort. Trying to capture such global variables as characteristics of people, workplaces, and work processes into three measurement devices probably seems most absurd. However, utilizing elements related to the organizational model cited earlier, our team was able to classify the myriad unique characteristics into questionnaires that provide ways of assessing this fit. Results have not been wholly conclusive. We are, however, encouraged by the trends, and in my present capacity as corporate director of Health and Wellness for Herman Miller, Inc., I continue to work at refining both the instruments and our own thinking in order to produce procedures that will allow facility managers to assess the effectiveness of their activities vis-à-vis corporate profitability.

Our work in this area has been published in a book entitled *The Negotiable Environment: People, White-Collar Work, and the Office.* Briefly, what we have done is to quantify the variables that I have discussed as being the most relevant to managing physical workplaces—namely, the worker, the job that is accomplished by the worker, and the physical workplace. The manner in which we elected to measure the individual worker was through an instrument called the Myers-Briggs Type Indicator. This questionnaire consists of 126 items measuring an individual's preferences for perceiving the outside world and for making decisions about the information collected. The Myers-Briggs instrument is the most commonly used instrument to measure the typology of Carl G. Jung.

## People

The model for describing people is shown in Figure 8-2. This model, similar to the model presented for looking at cultures, is also based upon the intersection of two major axes. The horizontal axis represents the manner in which people prefer to perceive the world. On the extreme left of that axis is the variable called "sensing," which describes the process of becoming aware of things directly through the five senses and concentrating on the actual facts and details to be verified by seeing, hearing, touching, weigh-

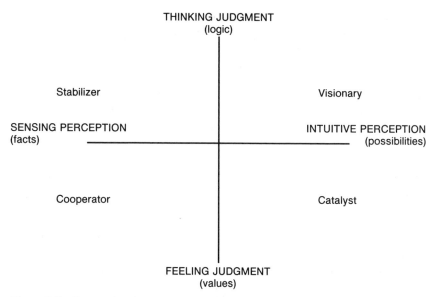

THINKING JUDGMENT
(logic)

Stabilizer

Visionary

SENSING PERCEPTION
(facts)

INTUITIVE PERCEPTION
(possibilities)

Cooperator

Catalyst

FEELING JUDGMENT
(values)

**Figure 8-2.** Personal styles as measured by the Myers-Briggs Type Indicator.

ing, and measuring. On the extreme right of the dimension is "intuition," which describes the process of becoming aware of things *indirectly* by incorporating ideas and associations from the unconscious with data provided through the senses. This process usually leads to a long-range, future-oriented approach to possibilities, hunches, and the desire for things to be better than they are. The vertical axis consists of the continuum from "thinking" to "feeling" and represents the extremes in the judging (decision-making) process. At the upper end of the axis is "thinking judgment," which describes the process of coming to a conclusion by logical analysis aimed at an impersonal finding. The opposite end of that dimension is "feeling judgment," which describes the process of coming to a conclusion by appreciating the personal, subjective values assigned to things, people, and events. Decisions made with feeling judgment are those that take into account subjective data.

The four types of persons described by this model are as follows: The *visionaries*, in the upper-right-hand quadrant, are the hard-driving achievers who value competence and logic in order to collect theoretical possibilities while subordinating human elements. They are people who strive to see "the big picture" in whatever they undertake and especially focus on the future and on dealing with the complicated interactions within organizations.

In contrast, persons who fall in the lower-right quadrant of the model are called *catalysts*. They gather information similarly to the way visionaries do but act on that information very differently. Catalysts' decisions are based on the implications for all people involved, including themselves, and are aimed at making everyone happy. They are people who manage by utilizing their personal charisma and tend to work better in democratically oriented or entrepreneurial organizations.

On the left side of the model are two types of persons who are more tactically oriented. In the upper-left-hand quadrant are the persons we have called the *stabilizers* because of combining "sensing" and "thinking" in the collection of hard data. Stabilizers tend to value both scientific and direct experiences and search for "truth" in developing procedures for action. Their orientation is to present procedures in a completely logical order, aimed at creating stability within an organization so that the organization can, essentially, run itself when the rules are followed. Somewhat in contrast, the persons in the lower-left-hand-quadrant (the *cooperators*) also like to deal with facts and direct experiences in light of the personal values of those affected by the information. Consequently, cooperators seek to have harmony with fellow employees. They tend to be practical in their approach to getting things done and will accomplish their work with enthusiasm and dispatch once they understand what is expected of them.

Let me point out that when comparing this model to the cultural model presented earlier in this chapter, we are most likely to find visionary managers functioning in the rational culture. Catalyst managers and other personnel are most likely to be found in the developmental culture. The cooperator, in the lower-left-hand corner of the personality model, matches the clan culture in Figure 8-1. Finally, the stabilizer in the personality model is most likely to gravitate toward and participate in the hierarchical culture described in the first part of this chapter. As a result, we are able to make comparisons between kinds of people and the cultures in which they work, making the assumption that similar types result in greater satisfactions.

## Work Process

The next step in the process of measuring the major ingredients of facility management was to develop a system for quantifying the work process actually accomplished by each individual worker. A significant amount of research has already been done utilizing the Myers-Briggs Type Indicator to identify different ways that the different personality styles have of approaching their occupations. We were able to draw upon that body of knowledge in constructing a matrix similar to the matrices utilized in Figures 8-1 and 8-2. Figure 8-3, however, is presented in order to convey some brief information about the differences in quadrants describing work processes found in our study. The instrument that was devised to measure the work process has the same dimensions as previously shown in Figure 8-2—that is, the horizontal dimension consists of sensing and intui-

THINKING JUDGMENT

| | |
|---|---|
| 1. Pays attention to today's concerns more than tomorrow's possibilities.<br>2. Develops detailed plans to guide actions for self and others at work.<br>3. Exercises good memory for details.<br>4. Performs many detailed concrete technical tasks.<br>5. Systematically follows an established procedure for accomplishing technical aspects of the job.<br><br>(General Business, Accounting, Manufacturing, Construction, Project Engineers) | 1. Takes pieces of information and translates it into broad categories.<br>2. Deals with nonroutine, technical activities such as dreaming up new uses for equipment, cutting through red tape, and so on.<br>3. Develops broad picture of work related situations.<br>4. Discovers connections among the various pieces of information associated with the job.<br><br>(Long-Range Planner, Systems Analysts, Research, Management, Physical Sciences) |

SENSING PERCEPTION ————————————————— INTUITIVE PERCEPTION

| | |
|---|---|
| 1. Deals with daily realities (facts and details).<br>2. Processes facts, information, etc. to determine if it supports or contradicts values and beliefs of self or company.<br>3. Uses company values to guide planning and scheduling even when this may produce some plans which appear illogical.<br>4. Does work which significantly affects personal attitudes and work lives of other people.<br><br>(Sales, Benefits Officers, Personnel, Health Care Workers, Community Liaison) | 1. "Plays by hunches" when concrete information is not available.<br>2. Communicates complex information to others.<br>3. Brings up all possibilities associated with a situation when working on job related problems.<br>4. Is guided more by general principles and concepts than by specific day-to-day demands.<br><br>(Training, Psychology, Communications Experts, Designers, Organizational Behavior Consultants) |

FEELING JUDGMENT

**Figure 8-3.** Sample work processes associated with psychological processes of decision making and perception.

tion; the vertical dimension of thinking and feeling. Figure 8-3 shows a few of the work items from the questionnaire that categorizes the work process among the four quadrants. Occupations that are most likely to be found employing the activities listed in each quadrant also are shown in Figure 8-3.

During different stages of their development, high-tech industries will require different activities to be accomplished. In the beginning, industry occupations in the upper-right-hand quadrant of Figure 8-3 are most likely to be prominent as new technologies and new organizational schemes are developed. Later, marketing and sales activities will be added. Similarly, as soon as a company begins production, it will need people performing duties from the upper-left-hand quadrant of the matrix in Figure 8-3. This measurement of work process completed two of the major dimensions that make up the person/place/process model of facility management.

## Workplace

The third, and final, dimension of this model is the measurement of the physical environment. Essentially, we ask "questions of the environment." The "answers," as recorded on a rating scale marked by an observer, identified the attributes of a workplace that might be characteristic of environmental "personality types." Seventy such attributes were considered. Examples are conference tables and chairs, visual barriers within work spaces, extent of personalization, extent of storage capacity, and location of the office in relation to traffic corridors. The rating scale distinguished, as much as possible, environmental "givens," thereby illuminating physical inhibitors, or enablers in the physical environment. The rating scale employed the same nomenclature as we used to measure cultural characteristics, people, and their jobs.

Samples of workplace characteristics are listed in Figure 8-4. As an example, workplace characteristics most likely to distinguish persons we have called visionaries (Figure 8-2, upper-right quadrant) are (1) ample storage for historical and reference material, (2) relatively high degree of privacy, (3) plenty of shelf space for books and documents, (4) work surface usually filled with work in progress, and (5) relatively few space constraints.

## The Fit

We studied 222 people from 12 companies with six grade levels. Three of those companies would be classified as high-technology workplaces. It is interesting to note that the distribution of personality types among the subjects corresponded almost exactly with that found among the general business population.

By contrast, however, we have found that the distribution of *work process* types did not match the distribution of personality types. The largest discrepancy occurred between the persons we have labeled stabilizers and their work processes. This suggests that since there are so many stabilizers in the business world (nearly twice as many as visionaries), many are doing work unsuited to their natural preferences.

THINKING JUDGMENT

| | |
|---|---|
| A work area geared toward action. | Ample storage for historical and reference material. |
| Easy access to procedures manuals, charts, graphic used in daily work. | Relatively high degree of privacy. |
| Desk and table tops used for work, not filing. | Plenty of shelf space for books and documents. |
| Evidence of electronic devices for analyzing and storing relevant data. | Work surface usually filled with work-in-process. |
| Very little evidence of personal effects or momentos. | Relatively few space constraints. |

SENSING PERCEPTION ———————————————— INTUITION PERCEPTION

| | |
|---|---|
| Chairs, tables, countertops for frequent use by visitors to the work area. | Presence of many personal items. |
| Less privacy required so may be located near center of activity. | Intermixing of work and personal items on bulletin boards and work surfaces. |
| Evidence of receipt and transfer of information such as "in" and "out" boxes. | Chairs and tables for use with visitors. |
| Work area neat and orderly in appearance. | Cluttered in appearance because desk and table tops used for "filing." |
| Evidence of working electronic devices. | Evidence of much unfiled but not used information. |

FEELING JUDGMENT

**Figure 8-4.** Physical workplace characteristics classified by decision making and perceptual styles.

Not surprising to us was the fact that the distribution of personality types also differed markedly from that of *workplace* types. In fact, this is the largest discrepancy: therefore, the source of the most *mismatch*. There are far more cooperator and catalyst offices than would be expected from the distribution of personality types. We strongly suspect that offices are far more likely to match the preferences of people who design and construct them (predominantly catalysts and cooperators) than those of the people who use them. We also discovered that offices are more likely to be designed for value-oriented ways of decision making (typical of developmental and clan types of organizations) than for the logical analytic mode (typical of hierarchical and rational organizational cultures) most prevalent in high technology.

In general, we found that new office environments seldom match the users that they must support. Little attention goes to accommodating personal work styles in office designs and construction. Far more effort goes to matching an office to the function it houses, regardless of the kind of person performing that function. Fitting an office to the work process is, of course, absolutely necessary. But that obviously is not the whole story.

## Comparisons with Other Organizational Measures

Since we have tried to examine the "whole story," we devised a scale to measure maximum matches and mismatches among the three elements that we measured—people, process, and place. We compared the best matches among those three and the worst mismatches with other measures of behavior: job satisfaction, satisfaction with present workplace, occupational satisfaction, individual rating of job performance, company satisfaction, and supervisor rating of job performance.

Persons who had the highest *matches* between personality, work process, and workplace also obtained higher scores on job satisfaction, company satisfaction, and satisfaction with present workplace. Our conclusion was that quality of *fit* between people, their work, and their workplace is, then, a measurable construct, definitely related to factors affecting individual, and ultimately corporate, productivity.

Only 7% of the total sample work in offices that coincide both with their work and their personalities. Those persons most likely to be well matched with respect to their personal characteristics, the work processes, and their physical places are engineers, accountants, programmers, and people who deal with things in specific, concrete, linear ways. Since many such occupations are found in high-tech industries, it is suggested that there may be somewhat fewer people who experience mismatching in high-tech industries than there are among employees of nonmanufacturing industries.

But on the other hand, those persons who are least satisfied with their work environment tend to be those classified as visionary existing in rational cultures. They are the persons in high-tech industries found in research departments, management in general, systems analysts, and the scientists connected to the organization. Both the catalyst and the cooperator are found some place between the two extremes.

## CONCLUSION

In conclusion, companies are not providing facilities that match both the people who are hired to do the jobs and the jobs that are to be done in those places. High-technology companies may in fact be providing facilities that are more consistent with the engineering and manufacturing parts of the corporation than with the research and development activities that must go on to provide long-term help for those industries. In general, there does not seem to be a good match between person, place, and process. Management consultants are predicting the decline of large hierarchical organizations because of the inflexibility in those systems in responding to technological changes. Our buildings must begin to reflect those kinds of changes in the corporate culture and the resultant changes in the kinds of people who are employed in the high-technology industries and the work that is required of those people. Careful attention to those variables will provide the information architects, planners, and builders need to design effective high-technology environments.

My attempt has been to demonstrate that high-technology environments are not all alike. They differ in their corporate cultures. Their work forces differ. The individual jobs differ. Therefore, their environments should match those differences. I presented evidence from work at the Facility Management Institute to demonstrate that there are more mismatches than matches in the companies we studied. Our task is to consider the design of high-technology work environments as a *management* issue rather than a technical problem in order to ensure having environments in which all parts work together for the accomplishment of corporate goals.

# BIBLIOGRAPHY

Deal, Terrence E., and Allan A. Kennedy. *Corporate Cultures: The Rites and Rituals of Corporate Life.* Reading, Mass.: Addison-Wesley Publishing Company, 1982.

Jung, Carl G. *Psychological Types.* London: Routledge and Kegan Paul, 1982.

Keegan, Warren J. *Judgements, Choices and Decisions: Effective Management Through Self-Knowledge.* New York: John Wiley & Sons, 1984.

Kimberly, John R., and Robert E. Quinn. *Managing Organizational Transitions.* Homewood, Ill.: Richard D. Irwin, Inc., 1984.

Mitroff, Ian I., and Ralph Kilmann. "Stories Managers Tell: A New Tool for Organizational Problem Solving." *Management Review* (July 1975: 18-28).

Myers, Isabel Briggs, and Mary H. McCalley. *Manual: A Guide to the Development and Use of the Myers-Briggs Type Indicator.* Palo Alto, Calif.: Consulting Psychologists Press, 1985.

Ouchi, W. G. *Theory Z: How American Business Can Meet The Japanese Challenge.* Reading, Mass.: Addison-Wesley, 1981.

Williams, Cecil L., David Armstrong, and Clark Malcolm. *The Negotiable Environment: People, White-Collar Work, and the Office.* Ann Arbor, Mich.: Facility Management Institute, 1985.

CHAPTER NINE

# ERGONOMICS IN HIGH-TECHNOLOGY WORK ENVIRONMENTS

## *T. J. SPRINGER*

**D**iscussions of high technology as it affects work and workplaces tend to focus on the capabilities, costs, and requirements of devices, both hardware and software. Yet the single most important element of the workplace system—which also happens to be the most valuable, adaptable, and unpredictable factor—is the human user. The scientific discipline that focuses on how human behavior shapes and is shaped by design is called *ergonomics*. In the United States, one often encounters the terms *human factors, human factors psychology, human engineering,* and *human factors engineering*. While there are fine points of distinction between these terms, for purposes of this discussion they will be treated as equivalent and the term *ergonomics* will be used.

The word *ergonomics* is taken from two Greek roots: *ergo*, meaning work, and *nomos*, meaning rules or laws. At the simplest level, ergonomics is the study of people at work. Technically, ergonomics is not a design science. Rather it provides information with which the design sciences can better accommodate the needs of people. Ergonomics evaluates how the things people use and the ways and places in which they are used affect people and how the characteristics of people can be used as input to the design process. Unlike other disciplines that shape the design of work and work environments, ergonomics is *anthropocentric* (human-centered) in its approach. This means ergonomics treats the human element as the primary determinant of workplace design. Data that describe people in terms of their physical, physiological, psychological, social, and cultural characteristics are used to define requirements for the design of tasks, tools, equipment, and environments.

Traditional workplaces have usually been designed from the outside in. Outside-in design starts with the building and moves progressively toward smaller delineations of space until the individual workstation is arrived at. This approach typically fosters designs based on a given amount of space into which the maximum number of people must fit. It also leads to workplaces based on space criteria rather than task or user requirements.

Ergonomics reverses this approach in specifying the operators' and operational requirements for workplaces. The fundamental premise of this technique is the belief that the worker and the task activities are critical determinants of space and support requirements.

## SYSTEMS APPROACH

When where and how people work are dealt with, it becomes apparent that the workplace is a complex system that affects human behavior in a great number of ways. Ergonomics utilizes a systems approach to problem solving. As in any complex system, there exists a level of synergy wherein the whole is greater than the sum of its parts; however, to begin to understand the system first it is necessary to describe the individual elements. In broad terms there are four critical elements in the workplace system. They are:

1. *People*—all the workers who occupy the workplace

2. *Tasks*—what people do and how they accomplish their job duties

3. *Tools*—the technologies, job aids, and equipment used in the performance of their jobs

4. *Facilities*—all aspects of the physical work environment

### Defining the User Population

Since ergonomics is an anthropocentric (human-centered) discipline, the first step is the definition of the population of users. When people are considered, it is apparent that there is a wide variety of shapes, sizes, abilities, backgrounds, etc. In fact, the only generalization that can be made is that everyone is different.

While people are extremely adaptable, they are relatively difficult to change (e.g., to make taller or younger). Consequently ergonomics focuses on defining the "envelope" of requirements of the human operator in order to adapt the work, workplace, and working environment to the person rather than vice versa. There are several ways in which this is done.

If one only envisions a group of people, it becomes apparent that no two humans are exactly alike. Among the more obvious differences is the variety of shapes and sizes of people. The description and measurement of body dimensions is called *anthropometrics*. Anthropometric data are generally collected for large groups of people. The large sample sizes characteristic of anthropometric measurement studies lead to the assumption that variations in body dimensions are distributed normally (the so-called bell curve). These data are reported as averages and summarized in tables showing values of a specific dimension at various percentiles of the population (Table 9-1). Thus ergonomic literature will talk of a 5th percentile (i.e., small) female or a 99th percentile male. One other characteristic of normal distributions that bears on the use of anthropometric data is the fact that the mean (average), mode (most frequent), and median (central point) are all represented by the value for the 50th percentile. However, in actual practice measures may not be distributed normally across the population or within the group of interest. Also, the data reported in anthropometric tables are often based on measures of select groups within the population, such as military personnel, and therefore do not reflect the complete range of variation that may exist in the population at large. As a result, use of anthropometric data must be tempered with an understanding of their inherent strengths and limitations and the way in which they are collected and reported.

A related source of information of importance in defining the population are biomechanic data. While anthropometrics deals with the body in static positions, *biomechanic* data relate to the human body in motion. Measures of strength, range of motion, and reach envelopes at particular angles from a reference point are typical of biomechanical measures. Biomechanical data are expressed in terms of measures for particular anthropometric dimensions. For example, reach envelopes at various angles above a seat reference point are reported for a series of different-size people as represented by percentiles in the population distribution (e.g., 1st, 5th, 50th percentile) (Table 9-2).

Numerous sources of anthropometric and biomechanical data are available (see References cited); however, the *Anthropometric Source Book Volumes I-III* (NASA 1978) is noteworthy in that it includes brief descriptions of the samples and sample sizes that go into the composite population data (1). *Humanscale 1-9* and *Human Dimension & Interior Space* are also good design references (2,3).

# Using Anthropometry

Anthropometric and biomechanic data are important in defining the population of users; however, the way in which these data are used in the process of design is critical to the success and acceptance of the end product. There are, in effect, three design approaches when using these types of data.

## DESIGN FOR THE EXTREME

In certain circumstances, the limiting factor of a design is the extreme of the population of users. For example, if in considering doorways one designs the height to accommodate the tallest person, everyone else will be able to pass through the doorway without difficulty. Similarly, if it is critical that an equipment operator be able to reach an emergency control, that control should be placed within reach of the smallest person, thereby being easily reachable by the remainder of the population.

## DESIGN FOR THE AVERAGE

When anthropometric data are used it is a common practice to design for the dimensions of the 50th percentile, or statistical average in a normal distribution. The theory behind this practice is that design for the 50th percentile requires the least accommodation from the largest group of people. In certain instances this is appropriate. For example, the counter height for public places such as grocery checkouts might be designed for the average person since designing for the extreme is inappropriate and adjustment is impractical. However, design for the average is not appropriate when "goodness of fit" is critical. Most people have encountered

TABLE 9-1
**Anthropometric measurements**

| MEASUREMENT NUMBER | DESCRIPTION | MINIMUM | | MAXIMUM | | DIFFERENCE | |
|---|---|---|---|---|---|---|---|
| | | IN | CM | IN | CM | IN | CM |
| 32 | Acromion to dactylion (length of the human arm) | 26.1 | 66.2 | 33.1 | 84.0 | 7.0 | 17.8 |
| 80 | Arm reach from wall | 27.9 | 70.8 | 38.5 | 97.9 | 10.6 | 27.1 |
| 194 | Buttock–Knee length | 18.2 | 46.3 | 26.9 | 68.3 | 8.7 | 22.0 |
| 200 | Buttock–Popliteal length | 14.8 | 37.7 | 22.8 | 57.9 | 8.0 | 20.2 |
| 312 | Elbow rest height | 6.1 | 15.4 | 12.6 | 32.0 | 6.5 | 16.6 |
| 330 | Eye height, sitting | 25.9 | 65.7 | 34.9 | 88.7 | 9.0 | 23.0 |
| 381 | Forearm–Hand length | 14.7 | 37.4 | 21.2 | 53.8 | 6.5 | 16.4 |
| 529 | Knee height, sitting | 16.6 | 42.2 | 24.6 | 62.4 | 8.0 | 20.2 |
| 572 | Maximum reach from wall | 30.5 | 77.4 | 43.1 | 109.5 | 12.6 | 32.1 |
| 612 | Midshoulder height, sitting | 20.6 | 52.3 | 28.0 | 71.1 | 7.4 | 18.8 |
| 666 | Patella top height | 15.2 | 38.6 | 24.1 | 61.1 | 8.9 | 22.5 |

TABLE 9-1 (Continued)
**Anthropometric measurements**

| MEASUREMENT | | MINIMUM | | MAXIMUM | | DIFFERENCE | |
| NUMBER | DESCRIPTION | IN | CM | IN | CM | IN | CM |
|---|---|---|---|---|---|---|---|
| 678 | Popliteal height | 13.1 | 33.4 | 20.2 | 51.4 | 7.1 | 18.0 |
| 758 | Sitting height | 29.6 | 75.2 | 39.7 | 100.8 | 10.1 | 25.6 |
| 867 | Thumb tip reach | 24.5 | 62.2 | 37.2 | 94.6 | 12.7 | 32.4 |

*Source:* NASA Reference Publication 1024, Washington, DC. Nasa Scientific & Technical Information Office, 1978.

products that claim to be universally applicable—"one size fits all" (e.g., clothing). But very few products can fit everyone. Why is it then that many workplaces and buildings have been designed with this principle in mind?

When considering more than several anthropometric measures, one must realize that there is no truly average person. In other words, a number of people may be of average height but within that subgroup there will be considerable variation in weight, leg length, arm length, etc. Conse-

TABLE 9-2
**Shirt-sleeved grasping reach: horizontal boundaries, 20-in. level**

| ANGLE (deg) | N | MIN | PERCENTILES (IN.) 5th | 50th | 95th |
|---|---|---|---|---|---|
| L165 | | | | | |
| L150 | | | | | |
| L135 | | | | | |
| L120 | | | | | |

| ANGLE (deg) | N | MIN | 5th | 50th | 95th |
|---|---|---|---|---|---|
| L105 | | | | | |
| L 90 | 11 | | | 14.00 | 18.75 |
| L 75 | 16 | | | 18.00 | 21.50 |
| L 60 | 20 | 17.00 | 17.50 | 20.50 | 24.50 |
| L 45 | 20 | 18.25 | 19.50 | 22.75 | 26.75 |
| L 30 | 20 | 20.25 | 21.50 | 24.75 | 28.25 |
| L 15 | 20 | 22.50 | 23.50 | 26.75 | 29.75 |
| 0 | 20 | 25.00 | 25.50 | 28.75 | 31.75 |
| R 15 | 20 | 27.25 | 28.00 | 30.50 | 34.00 |
| R 30 | 20 | 29.00 | 30.00 | 32.00 | 35.75 |
| R 45 | 20 | 30.50 | 31.00 | 33.50 | 36.25 |
| R 60 | 20 | 31.50 | 32.00 | 33.75 | 36.25 |
| R 75 | 20 | 31.50 | 32.25 | 34.00 | 36.50 |
| R 90 | 20 | 31.75 | 32.25 | 34.00 | 36.00 |
| R105 | 20 | 31.50 | 31.75 | 33.50 | 35.75 |
| R120 | 19 | | 30.50 | 33.00 | 35.50 |

TABLE 9-2 (Continued)
**Shirt-sleeved grasping reach: horizontal boundaries, 20-in. level**

| ANGLE | | PERCENTILES (IN.) | | | |
|---|---|---|---|---|---|
| *(deg)* | *N* | *MIN* | *5th* | *50th* | *95th* |
| R135 | 9 | | | | 34.50 |
| R150 | | | | | |
| R165 | | | | | |
| 180 | | | | | |

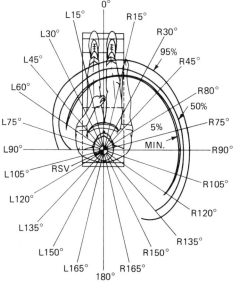

Source: H. P. Van Con & R. G. Kinkade. "Human Engineering Guide to Equipment Design." New York, McGraw-Hill, 1963, p. 530.

quently by designing for the statistical average "person," one is in reality designing for a nonexistent user.

The fallacy of the average person was demonstrated dramatically by G. S. Daniels in his study "The Average Man?" (4). Based on anthropometric data collected from over 4,000 U.S. Air Force male flight personnel, the study examined 10 body dimensions to see how many of the sample population were average. Rather than utilize exact averages, Daniels allowed a range of plus or minus 15% on either side of the exact average. This 30% range for each dimension was called the approximate average. In stepping through the 10 measures, none of the original 4,063 men were average on all measures. In fact, after the fourth step less than 2% of the population remained and any significant correlation had disappeared.

## DESIGN FOR THE ADJUSTABLE RANGE

Design for the adjustable range recognizes the inherent variation within the population and attempts to accommodate the largest number of users by adjustments in the product. It is common to select a range from anthropometric tables that a design attempts to accommodate (e.g., 10th to 90th, 5th to 95th, 2.5 to 97.5, 1st to 99th percentile). A common range is 5th to 95th percentile for a given population. Since people vary considerably across various body dimensions, it is not uncommon for an individual to fall within the 20th percentile for stature and below the 5th percentile for popliteal height (i.e., height to underside of knee in seated position). Thus, choosing to exclude 10% of the population by selecting a range of 5th to 95th percentile on a particular anthropometric criteria will in fact result in the exclusion of *more* than 10% of the population. Consequently, a more conservative approach is strongly recommended through the use of a broader range of the population on selected measures (e.g. 2.5th to 97.5th or 1st to 99th percentile).

Whenever possible or practical, design for the adjustable range is preferable since this approach attempts to adapt the device to the user rather than vice versa. In design of workplaces, seating has been an area for which this design approach has been adopted. The adaptation of gas cylinders used originally in the auto industry has led to the design of chairs that are easily adjusted in height and, depending on the model, several other axes of adjustment. However it must be noted that when designing adjustment into workplace products, one must consider two types of adjustment: adjustment of the device to fit the physical requirements of the user (e.g., size) and adjustment to accommodate the performance of the task (e.g., task postures). Among products designed for use by the general population, automobiles offer excellent examples of this design approach. Six-, eight-, and 10-way adjustable drivers' seats, tilt steering wheels, and other adjustment features are provided to accommodate the vast range of sizes among the driving population.

Once the dimensional requirements of the user group are defined and the design approach selected, other data are important to specify user requirements. Among these are physiological data. Physiological data relate to the human being as a biological organism. In practical terms, physiological data are used to set health and safety criteria for environmental elements within workplaces. These data are often expressed as safety standards or *threshold limit values* (TLV). People require certain environmental conditions to survive. A breathable atmosphere, certain temperature limitations, protection from radiation, etc., normally are considerations only when one is designing work for a "hostile" environment. Some of the best examples of how these considerations shape design for a hostile environment are found in the space program. However, assuming for the time being that the majority of human beings will remain earthbound, we can accommodate minimal survival requirements by natural environmental elements. Thus, attention can be focused on the relatively narrow environmental envelope required to support safe and healthy work activities. Ways in which these requirements are met can determine design parameters of equipment and workplaces. For example, in extremely cold environments, protective clothing will require that knobs, handles, and the like allow use by gloved hands. Another common example is air quality in modern workplaces. Of importance in determining minimal air quality standards are the amount of airborne material, the rate of air exchange, etc., all of which can be referenced to human requirements for fresh air, temperature control, humidity, etc. Similarly, safety signals in an extremely noisy environment might require multiple cues (such as flashing lights, loud sirens, and vibrations) to compensate for the high level of ambient noise and the need for workers to wear ear protection.

A vast amount of behavioral science data exists that describes how people respond and react to physical, psychological, social, and cultural events or stimuli. Sensation and perception data that describe people's ability to sense and understand physical stimuli are the behavioral extensions of physiological data. For example, noise levels well below that which can damage hearing nevertheless can be extremely irritating and lead to behavioral changes among workers. Similarly, certain colors of light change the way colors appear. While the illumination level is well within acceptable range for sight, the effect on perceptions makes their use unacceptable.

Sociocultural data are important in defining such things as needs for privacy, personal space, effects of stress, and affiliative need. Sociocultural data also deal with stereotypical behaviors among the target population. For example, in the United States automobiles are driven on the right side of a road; traffic signals are most often arranged as a vertical, (red [stop], yellow [caution], green [go]) multilight display; doorknobs are turned; buttons are pushed. All of these are designs that through widespread use have instilled common stereotypical behaviors in a population. Violation of these instilled behavioral stereotypes at this point in time will be resisted and can lead to errors or alteration or rejection of the design. All of which demonstrates that while design affects behavior, behavior should also affect design.

These last two types of data are important in an emerging area termed *cognitive ergonomics*. This branch of ergonomics deals with the ways in which the mediation aspects of human/technology interaction are designed to accommodate how people think and behave. For example, in the design of human computer interfaces, an area of major concern and activity for cognitive ergonomics is structuring software systems so that they are easy for neophyte users to learn yet are flexible enough to be used effectively by both novices and experts. Such flexible systems provide aid for those who need it but do not penalize expert users by forcing them to scroll through what are, for them, unnecessary tutorial screens. Similarly, in designing jobs that include computers, cognitive ergonomists might address such issues as what people can do when computer system response time is slow or when the system is "down." One final example of the growing importance of cognitive psychology and ergonomics in high-technology product design is in *intelligent systems*. So-called *expert system* software products are designed to provide access to encyclopedic information on specific topics. Input from cognitive psychology is critical to successful design of such products since the way in which information is organized internally, the logic patterns used to identify possible solutions and the format used to pose questions and present information not only must follow human thought processes but also must be nonthreatening in order to be accepted as a useful tool. Not to belabor the point, but this is yet another area in which behavior is used to shape design and design to shape behavior.

## Describing the Work

Once basic descriptions of the worker population are known, the next question to be addressed is, What do these people do? For purposes of ergonomics, describing the processes of work differs somewhat from the traditional personnel approach to job description. Often what people do departs considerably from the formal duties listed in their job descriptions. *Systems analysis* of work involves observing actual work behavior and breaking the activity into discrete components (i.e., tasks). The ergonomist focuses on any impediments to performance, on necessary training, and on jobs aids required to complete the task successfully.

Systems analysis begins with a step called *functional requirements definition*. Functional requirements definition relies upon an accurate description of what must be done to accomplish the stated goal of the system (e.g., produce microprocessor chips). The discrete steps are then diagrammed in block fashion. *Functional block diagrams* can be of progressively greater detail and involve an analysis of human and equipment capabilities. Once the tasks and functions have been described, the ergonomist looks for those activities that can be best accomplished by the human element and those best performed by a tool, technology, or piece of equipment. This step, called *function allocation*, is extremely important because it allows the ergonomics professional to weigh the advantages and disadvantages of using either humans or machines to accomplish each of the system functions. An important consideration in this process is the cost-effectiveness of one alternative versus the other. Knowledge of the capabilities and limitations of the worker population gained from the above data is critical to this process.

An important component of systems analysis is mapping the flow of information through the system. This can be done informally as part of the function block diagram process or in a more formal fashion by charting the information flow separately. Like functional block diagrams, *information flow charts* can contain increasing degrees of detail. Functional information flow charts are very similar to the flow charts developed by computer programmer analysts. At each step in the information flow chart, operations and decisions required to accomplish a given function within the system are reduced to binary choices (e.g., yes/no; on/off). By stridently reducing operations and decisions to binary choices, the flow chart can become a usable model for computer programming of certain system elements as well as an easily understood map of how information flows through the system.

Following the identification and allocation of functions and the analysis of information flow, the next step is analyzing the tasks. *Task analysis* can be very general or extremely specific; there are no set rules. However, there are two key purposes of task analysis: first, to evaluate what is required of each of the workplace elements to accomplish the given task or sequence of tasks; second, to identify ways in which performance of the task can be optimized from the human standpoint. Techniques include *motion analysis* and *link analysis*. Motion analysis maps the discrete motions and corresponding times to perform the activities that constitute a task or series of tasks. Link analysis is used to optimize arrangements of elements. In link analysis, motion analysis data are often used to plot patterns and frequencies of actions required to complete a particular task *sequence.* Configurations of workplace elements are then optimized by reducing motion times and distances for the most frequent or critical activities. Link analysis has been used extensively to improve the layout of display panels such as those used in military aircraft or process control workstations (e.g., nuclear power plants). However, it is equally applicable in determining other "optimal arrangements"—for example, machines on an assembly line, minimal distances among workers (similar to adjacency analysis), or configurations of equipment within a workstation. The term link analysis comes from the description of connections between people and equipment or people and other people. Common types of links are auditory, visual, or control in nature; however, there may be other types of links unique to specific systems. Following identification and description, the links are weighted according to their importance in performing a given job. Through the process of describing and weighting task sequences and links among activities, the ergonomist can develop ways to remove impediments to job performance and to streamline operations.

These traditional techniques of systems analysis tend to concentrate on observable activity (i.e., motion). However, as the handling of information plays an increasing role in all jobs, mediation, decision making, and other unobservable behaviors increase. Consequently new techniques of analyzing job behaviors must emerge. One such instrument is the Position Analysis Questionnaire (PAQ). Developed by McCormick, Jeanneret, and Mecham (5), the PAQ consists of 189 work-oriented job elements making up six major divisions of worker activity (Table 9-3). Note that physical work activity is but one of the major subdivisions.

Although the definition of tasks and task behavior involves an understanding of the tools involved and how they are used, the principal concentration is on the goals of the work. For example, the general task of data entry may be very similar whether one is using a keypunch on a computer terminal connected to a mainframe or a personal computer (PC), while the actual devices are considerably different from one another.

Since the goal of ergonomics is to provide work environments that are safe, comfortable, easy to use, and productive, of equal importance to what tasks are performed is what tools, technology, and equipment are required and how they are used in the performance of tasks.

---

TABLE 9-3
**Major subdivisions of worker activity from the position analysis questionnaire (PAQ)**

1. Information input
   Sources of job information

2. Mediation Processes
   Decision making and reasoning
   Information processing
   Use of stored information

3. Work Output
   Use of physical devices
   Integrative manual activities
   General body activities
   Manipulation coordination activities

4. Interpersonal Activities
   Communications
   Interpersonal relationships
   Personal contacts
   Supervision and coordination

5. Work situation and job context
   Physical working conditions
   Psychological and sociological aspects

6. Miscellaneous aspects
   Work schedule, method of pay, and apparel
   Job demands
   Responsibility

*Source:* F. J. Landy and D. A. Trumbo (1980). Psychology of Work Behavior. Homewood, Ill., The Dorsey Press, 1980, p. 107, by permission of Brooks/Cole Publishing Company.

One of the things that distinguishes humans from other species is our ability to make and use tools. Since the first prehistoric toolmaker notched a flint hand ax to accommodate a better grip, people have striven both intuitively and through applied science to improve the apparatus used in performing tasks. Technology is simply the application of millions of years of toolmaking experience and expertise backed by the scientific knowledge gained through controlled experimentation. As people use technology they attempt to discover ways to adapt the tools to their needs, often using technology in innovative fashions far beyond the intent of the original design. What ergonomics strives to do is identify the work-related requirements of people and tasks and provide input into the design of tools and training of users to maximize the interface between people and technology.

The techniques used in evaluating tool and technology requirements build on the results of the preceding analyses. By performing function analysis, function allocation, information flow analysis, and task analysis, the ergonomics professional develops an understanding of the task requirements, the limitations of the human operator, and those task elements that are best performed by technology. From these results a model of the human/technology system can be developed. Critical to this model are the data describing the capabilities, limitations, and expected behaviors of the population of users. Once a model is developed it is tested, first on paper and subsequently in mock-up and functional prototype form. Finally, through repeated testing and feedback, the design of tasks, tools, technology, and support equipment are shaped to accommodate the health, safety, comfort, ease of use, and performance of the human operator, and in turn system performance is optimized.

## Describing the Work Environment

The final question to be addressed by the systems approach deals with the workplace facility. As such it deals with issues of lighting and illumination, HVAC air quality, acoustics, power, cabling, space, aesthetics, etc. All of the physical elements of the work environment. Considerable data exist to guide architects and designers in the programming and planning of work environments; however, it is true that the modern manifestations of the architectural tradition have often forgotten the fact that buildings are meant to be not only artistic but also occupied. Consequently many cities sport "designer label" buildings—those noteworthy structures that are easily identified as having been designed by the latest "wunderkind" whose style is so recognizable as to be more art or fashion than an application of the time-honored form follows function approach. It is not surprising that in an era in which people proudly display the name or signature of clothing designers on their backsides, we should find the trend continues to structures; however, unlike clothing fashion, buildings are difficult to change and must be "worn" by everyone who occupies the space. Those responsible for the design of buildings of today and tomorrow must recall that, in the tradition of Le Corbusier, buildings are machines for living and working and to be successful they must be human referenced.

Starting with the individual workstation and expanding throughout the organization, the requirements of the worker, the task, and the technology must define and shape the specification of facility systems. For example, in an office building, cable requirements can vary tremendously depending on whether a space will be used for file storage, data entry, or telemarketing. It is a general trend that those jobs that require extensive use of files or reference materials require more floor space than those jobs that handle a minimum of such information. Consequently the amount of space and possibly the load-bearing characteristics of floors designed for legal, tax, or computer programming may be different from those designed for general clerical or management. Conversely the acoustical privacy requirements of managers are considerably different from those of data entry personnel. Cooling requirements for environments that include a concentration of electronic devices (e.g., PCs) are far greater than they would be for the same workplace without the electronics. Window treatments in a building that must accommodate visual display terminals (VDTs) will necessarily be different from those for a workplace without VDTs. The point here is that in an optimal system, the definition of user, task, and equipment requirements will define to a great extent the facility specifications.

## SYSTEM SYNERGY

Once the basic elements of the workplace system have been identified and described, the ergonomist begins to map interactions and interfaces within the system. Interactions are those regions of overlapping influence. Those areas where interactions occur are the interfaces, the regions of information exchange. The primary interest of ergonomics is the interface between the human element and the remainder of the system.

People are the most adaptive element in the workplace model. Consequently, much of the accommodation that has been demanded in nonoptimal situations has been demanded of the workers. Unfortunately, excessive demands placed on workers can lead to lowered productivity, errors, fatigue, stress, injury, and disease. The inappropriate match between existing environments, new technologies, altered jobs, and skilled workers highlights the growing need for ergonomics in the workplace. Without careful planning and attention to changes in requirements imposed by new technologies, situations arise in which the composite elements of the workplace system work against one another. One of the first symptoms of dissonance in the high-technology work environment is increased employee distress. It is not difficult to find workplaces where workers complain of eye strain, back aches, fatigue, numbness in limbs, headaches, tension and a myriad of other discomforts. Often physical or psychological discomfort has its roots in the characteristics of the designed elements of the workplace.

## THE EXAMPLE OF THE AUTOMATED OFFICE

If one were to examine the traditional office workplace, it would be apparent that the outside-in approach to design prevailed. Buildings were constructed, spaces were planned, and facilities were furnished with little consideration of the support needs of work or workers. Until very recently, this was due to the relatively low cost of labor, the high cost of buildings and equipment, and the fairly unchanging nature of office work and equipment. Except for some minor cosmetic alterations, the office changed very little between 1750 and 1950. Of course, some modernization of tools occurred; however, the majority of traditional office tools were originally introduced prior to 1900 (Table 9-4). Similarly, office work remained stable during this time period. The principal tasks of collecting, recording, storing, retrieving, and communicating information remained the same. The ways in which these were accomplished relied on the medium of information—paper. Consequently, the tools and equipment, and to a certain extent the environment, of the office evolved to serve the handling and flow of paper through the organization.

In this century, the office became a reflection of the factory with industrial techniques applied to office work and workplaces. Thus, work segmentation, equipment standardization, and "assembly line" work flow planning came into vogue in the office. Similarly, environmental elements were affected by the drive toward standardization. Lighting, for example, was uniformly bright, to facilitate reading "the fine print." In the era of cheap energy, it was not uncommon to find offices where the ambient illumination level was more than twice that required to comfortably read paper-based information. Similarly, desks and chairs for the most part came in one standard size, were arranged in rows and columns, and varied little across jobs or individuals. Much traditional steel office furniture appeared to be designed to be easily manufactured, arranged, and maintained rather than to meet the task or comfort requirements of the workers. Examination of dimensions of traditional office furniture suggests that the products were designed to fit the average adult male. Where adjustments were possible, they were difficult to make and often required the use of tools and the aid of maintenance personnel.

**TABLE 9-4**
**A chronology of certain technologies affecting office evolution**

| YEAR | INVENTION | TYPE |
|------|-----------|------|
| 1700 | | Manual |
| 1785 | Bifocal eyeglasses | |
| | | |
| 1800 | | |
| 1850 | Roll-top desk | |
| 1858 | Lead pencil | |
| 1867 | Typewriter | Mechanical electric |
| 1876 | Telephone | |
| | Mimeograph | |
| 1877 | Voice recording (wax cylinder) | |
| 1878 | Cathode ray tube | |
| 1879 | Commercial indoor lighting | |
| 1884 | Fountain pen | |
| 1887 | Graphaphone (wax cylinder dictation) | |
| 1888 | Ball-point pen | |
| | Adding machine | |
| 1899 | Magnetic tape voice recording | |
| | | |
| 1900 | | |
| 1906 | Photostatic document reproduction | |
| 1908 | Indirect incandescent lighting | |
| 1922 | Microfilm information storage | |
| 1938 | Commercial fluorescent lighting | |
| 1939 | Xerography | |
| 1944 | Automatic computing device (U.S. Navy) | |
| 1951 | Commercial computer (UNIVAC) | |
| | Commercial automatic xerography | |
| 1952 | Magnetic tape-loop dictation device | |
| 1956 | Electric portable typewriter | |
| 1958 | Solid state electronic computer | Electronic |
| 1960 | Electronic computer-thin film memory | |
| 1964 | Magnetic tape word processing | |
| 1972 | Video display (CRT) word processing | |

Only recently has white-collar, information work become an industry unto itself. Until the 1970s the primary medium of information was paper; however, with the introduction of the CRT-based word-processing systems of the 1970s electronics became the new medium of information. Of the many new technologies that have found their way into the workplace, few are as radically different from their predecessors as the VDTs in the office. Since paper will not disappear anytime soon, one of the challenges of designing for modern offices is the need to accommodate the use of both electronic and hard copy displays. The difficulty of doing so efffectively is illustrated by the characteristics of the information displays (Table 9-5) that impose substantially different requirements on the user, the support equipment, and the work environment.

**TABLE 9-5**
**Characteristics of display media**

| HARDCOPY/PAPER | ELECTRONICS/VDT |
| --- | --- |
| Dark characters | Luminescent characters |
| Light background | Dark background |
| Continuous line characters | Dot-matrix characters |
| Flat, matte surface | Curved, reflective surface |
| Horizontal plane | Vertical plane |
| Manual input | Keyboard input |
| Easily manipulated | Difficultly manipulated |
| Data stored physically | Data stored electronically |
| Perceptually permanent | Perceptually transient |
| Simultaneous presentation | Serial presentation |

If one examines these differences between paper and VDTs as displays, it is apparent that they impose almost opposite requirements on the user and the supporting workplace. Paper as an information display has evolved along with the equipment and environment to facilitate its use. Most paper used in the office is a light color, matte finish and has dark continuous line characters. These characteristics allow paper-based information to be read under a variety of lighting conditions and from a wide range of viewing angles. VDTs on the other hand, are light emitting as well as light reflecting. Information is presented through a curved, reflective glass surface. The characters are in a dot matrix and often are luminous characters on a dark background. The VDT image can be further deteriorated by dust on the screen or by veiling reflections from sources of illumination in the work environment.

Paper is normally handled in a horizontal plane (desk top) and can be easily manipulated manually. When writing is being done the eye and hand move in a coordinated fashion in which the eye follows the motions of the hand, and most of the weight of the arm is supported by the writing surface. By contrast, VDTs present information in a nearly vertical position. In spite of the introduction of the mouse, joystick, light pens, and touch screens, computer terminals still require the use of keyboards. Typing on a VDT keyboard is a complex psychomotor activity involving both hands, which must operate in a different visual plane from that of the information display. Operating a keyboard requires muscular support of the wrists and forearms involving the upper arms, neck, and back.

Except for extremely bulky reference materials, paper can be moved readily to positions where it can be more easily read. While VDTs are getting progressively smaller, they are still difficult to move easily and are connected by cabling to other devices. Consequently, users must adapt their position to the VDT rather than vice versa.

Paper is perceived as relatively permanent; it can be handled physically. Although paper may be misfiled or misplaced, perceptually people know it still exists. By contrast, electronic information displayed on a VDT is perceived as being very transient. One cannot handle electronic information physically. Power surges or interruptions, computer malfunctions, static electricity, magnetic fields, keystroke errors, and software bugs can all cause information to disappear. Thus how people think and feel about information in the two forms is very different and affects how people act and react to the media.

Information displayed on a VDT is limited by the size of the display area and the size of the characters. Information is often presented sequentially or serially on a number of "screens," through which the user must "scroll." Viewing information from more than one screen is difficult. While the amount of information presented on a particular piece of paper is also limited by its size, the actual size of paper-based information display is limited only by the physical size of workplace area and how many pieces of paper the user can spread out. Paper can present different items of information simultaneously, thus allowing easy comparison and rearranging of information from more than one source.

In light of the differences highlighted by the comparison of information media characteristics in Table 9-5, it is not surprising that the introduction of VDTs into the general working area of the office has given rise to a host of complaints from users. That is not to say that the underlying problems did not exist prior to the introduction of the VDT; however, the introduction of dramatically new technology and the concomitant change in support requirements imposed by the VDT magnifies the shortcomings of traditional office environments.

If the human/machine interface is the criteria around which the equipment and workplace environment are to be designed, the approach and result are considerably different from one in which the worker and the tools are expected to adapt to existing conditions.

## ECONOMICS OF ERGONOMICS

One of the challenges facing facilities managers today is the balancing of relative life cycles of the major elements of a workplace. Technology, specifically computer technology, is changing very rapidly, to the extent that new products are introduced every 18 to 36 months. Capital equipment such as furniture is usually tied to depreciation schedules of 5 to 7 years. Employees' effective careers last from 15 to 25 years. Buildings are in use anywhere from 25 to 75 years or more. Thus, not only must facilities managers match the capabilities and requirements of people, products, and places but they must deal with substantially different rates of change among these elements.

The relative expenses associated with the four basic elements of the workplace also vary considerably. For example, it has been estimated that personnel consume over 85% of costs associated with an office facility over a 10-year period. The remaining costs are divided among the building (7%), equipment (5%), and miscellaneous costs. Since people are the most expensive element by far and are rather difficult to change, both the workplace design and management must strive to maximize the value of personnel dollars by investing in those things that can be changed easily (i.e., the tasks, tools, and facilities).

In an examination of the impact of the physical work environment on productivity and performance, Springer (7) found empirical evidence demonstrating improvements of from 3% for elimination of glare to over 500% for changes in management, work, and workflow. In one study, moving from traditional, nonadjustable workstation furniture to adjustable, "ergonomic" furniture yielded an improvement of 10% for data entry tasks and 15% for a dialog task. These performance improvements translated into payback periods of less than 1 year (8). Clearly, the economic impact of ergonomics is significant.

## CONCLUSIONS

When high-technology environments are being designed, ergonomic considerations can be expressed in terms of four basic questions: Who are the users? What do they do? What is the role of technology? Where is the work performed? The answers to these questions will provide the information necessary to address the following concerns:

Accommodating a wide variety of uses and users

Integrating workers, work, workplaces

Providing adjustment in devices, support equipment, and workplaces

Providing effective facility support

Making effective use of space

One interesting phenomenon surrounding the science and application of ergonomics is the fact that the most effective solutions *appear* to be intuitive or even transparent to the user. Many feel ergonomics is simply applied common sense because designs that incorporate sound ergonomic principles feel right and make sense. Fortunately, the economic evidence proves that ergonomics makes dollars as well as sense. While Voltaire's observation that common sense is not so common is very true, it is the research-based, human-centered systems approach that distinguishes the science of ergonomics from simple common sense.

## REFERENCES CITED

1. Webb Associates, ed. *Anthropometric Source Book Volume I: Anthropometry for Designers; Volume II: A Handbook of Anthropometric Data; Volume III: Annotated Bibliography of Anthropometry*. NASA Reference Publication 1024. Washington, D.C.: National Aeronautics and Space Administration, Scientific and Technical Office, July 1978.
2. Diffrient, N., A. Tilley, and D. Harman. *Humanscale 1/2/3/4/5/6/ 7/8/9*. Henry Dreyfuss Associates, Cambridge, Mass.: MIT Press, 1981.
3. Panero, J. and M. Zelnik. *Human Dimension and Interior Space: A source book of design reference standards*. London: Whitney Library of Design, The Architectural Press, 1979.
4. Daniels, G. S. "The Average Man?" As quoted in Kleeman, W. "Flexible Interior Plug Into the Computer Age." *The Construction Specifier* (November 1982).
5. McCormick, E. J., P. Jeanneret, and R. C. Mecham. "A study of Job Characteristics and Job Dimensions as Based on the Position Analysis Questionnaires." *Journal of Applied Psychology*, 36:347–368, 1972.
6. Landy, F. J. and D. A. Trumbo. *Psychology of Work Behavior*. Homewood, Ill.: The Dorsey Press, 1980.
7. Springer, T. J. *Improving Productivity In The Workplace: Reports From the Field*. St.Charles, Ill.: Springer Associates, Inc., 1986.
8. Springer, T. J. *Visual Display Terminal Workstations: A Comparative Evaluation of Alternatives*. Bloomington, Ill.: State Farm Mutual Automobile Insurance Company, 1982.

## ADDITIONAL READING

Damon, A., H. W. Stoudt, and R. A. McFarland. *The Human Body in Equipment Design*. Cambridge, Mass.: Harvard University Press, 1986.
*Ergonomic Design For People At Work: Volume 1*. Eastman Kodak Company; The Human Factors Section; Health, Safety and Human Factors Laboratory. Belmont, Calif.: Lifetime Learning Publications, 1983.

*Ergonomic Design For People At Work: Volume 2.* Eastman Kodak Company; The Human Factors Section; Health, Safety and Human Factors Laboratory. Belmont, Calif.: Lifetime Learning Publications, 1986.

Grandjean E. *Fitting The Task To The Man: An ergonomic approach.* London: Taylor & Francis, 1982.

Kleeman, W. B., Jr. *The Challenge of Interior Design.* Boston: Science Books International, 1981.

Lueder, R., ed. *The Ergonomics Payoff: Designing the Electronic Office.* Toronto: Holt Reinhart & Winston of Canada, 1986.

Woodson, Wesley E. *Human Factors Design Handbook: Information and Guidelines for the Design of Systems, Facilities, Equipment and Products for Human Use.* New York: McGraw-Hill Book Company, 1981.

## CHAPTER TEN

# STANDARDS FOR OVERALL BUILDING PERFORMANCE, AND INFORMATION TECHNOLOGY

## *GERALD DAVIS*

### HOW TO KNOW WHETHER A BUILDING IS PERFORMING AS IT SHOULD

**M**aking buildings *work* is an extraordinarily complex task. Part of the complexity is due to the fact that so many people participate in planning and creating buildings. And after a building has been created, still other people, including the building occupants, have a role in making it work. Depending on their role and their prior training, they all have different objectives and expectations for what constitutes a building that works well.

### WORKS WELL *FOR WHOM?* MEASURED BY *WHAT CRITERIA?*

The people who might answer these questions will probably come from specific fields such as property development, finance, architecture, some form of engineering or one of the businesses or professions that occupies a building. What "works" for one is likely to be less satisfactory for others. A lighting fixture may deliver just the right amount of illumination according to the calculations of an engineer, but a space planner who later needs to arrange workstations in a layout unforeseen by the engineer may find the fixture quite unsuitable. Then, whether the light source would pass a test for output of illumination would be irrelevant to an occupant in the space.

## OVERALL PERFORMANCE IS WHAT COUNTS FOR THE OCCUPANTS

For the occupants using the space, what counts is *overall performance,* the *combined effect* of what the engineer and the space planner do. In high-technology environments, the complexity is compounded. Whether the occupants use high-precision instruments or video display units or clean rooms, they need all elements of the visual environment, *taken together,* to produce a specified level of glare-free illumination for whatever task they are working on.

An aspect of overall performance—such as suitability for people working at computer terminals or at workbenches for electronic testing—involves many interactive performance considerations that have previously been measured only separately. For instance, overall performance to meet functional needs of workers requires consideration of temperature, humidity, air quality and pollutants, seating, the range of available workstation dimensions and convenience of adjustment, the visual environment, acoustics and vibration, and so on. Overall performance to meet organizational requirements involves issues such as: cycles of use of the facility (diurnal, weekly, and annual); degree of control of the physical environment by occupants; extent of compartmentalization of the spaces; desired character and image; nature and extent of organizational change, now and in the future; and the style of facility management now and in the future. The equipment may function only when a whole array of individual performance requirements are met, such as those for space and arrangements, temperature, humidity, control of certain pollutants, control of dust, availability of cables for electrical power and data communications, control of magnetic fields, static, physical security and information security, and change in all these over time.

## STANDARDS CAN HELP

Standards make it easier to specify what type and quality of performance is required because they (1) incorporate the accumulated experience and best judgment of many people with relevant knowledge and interest, (2) reflect generally accepted practices, and (3) are written so anyone in the field can understand and comply. Standards save work because they permit the owner, occupant, or designer to simply specify a generally accepted performance requirement, instead of having to think through every aspect of how a building must perform.

There are standard criteria and tests for most of the parts and components of a building, such as lighting fixtures, doors, heating ducts, carpets, or materials such as paint, wood, steel, or concrete—that is, standards for what they must be made of or how hard or soft they must be. Compliance can be measured with tests that have been made "standard" by some process of agreement or approval. Standards also exist for some aspects of performance, such as how well an aluminum sliding door must keep out rain under certain test conditions or how well a paint surface must resist rubbing under certain laboratory conditions.

What have been lacking, however, are standards for *overall performance*—that is, how all the parts, components, systems, and materials of a building, taken together, meet a particular need of the occupants to be able to do something, accomplish a task, carry out a function or activity. A standard for measuring an aspect of overall performance considers how a whole building, or a whole space or facility within a building, meets a particular need of the occupants.

Standards for overall performance are important in many circumstances, such as: deciding on the purchase, lease, or occupancy of a facility; deciding whether to rehabilitate or renovate; giving instructions to those responsible for design or construction; and designing or purchasing the furniture and equipment to be used in a facility. Standards are important also in more contentious circumstances, such as litigation over a building, labor negotiations about working conditions, or labor grievances regarding the physical work environment.

## STANDARDS FOR OVERALL PERFORMANCE OF BUILDINGS ARE NEW

The need for standards for overall performance has been recognized only recently. Until the late 1950s, most designers specified the materials to be used in a building, such as the concrete, glass, paint, metals, acoustic tile, and sealants; and the components and systems to be provided, such as sliding aluminum doors, hung ceilings systems, and curtain wall systems. In the 1960s, the questions became broader: How should these materials, parts, and systems perform in actual use? For instance, the School Construction System Development project, led by Ezra Ehrenkrantz, specified that a wall panel in a high school must resist the impact of a football player with shoulder pads.

In the 1970s the distinction between functional and technical performance was made explicit. An occupant might state the functional requirements for a conference room as a place where 25 people can meet and understand each other without the need for electronic amplification and where they can show 35-mm slides, overhead transparencies, and videotapes. A project manager could translate this into a technical performance requirement against which design drawings and specifications would be evaluated and would specify criteria such as the sound transmission coefficient from or to adjacent spaces and the ambient sound level measured in decibels. The performance concept in building encompasses actual performance of a wall in place, as actually built, which often produces quite different evaluations than tests done in a laboratory of manufacturers' sample door assemblies and prototypical wall assemblies.

In the 1970s and 1980s, the specification of functional and technical performance requirements for facilities emerged in places as diverse as the U.S. Public Building Service, the U.S. National Bureau of Standards (Center for Building Technology), Building Officials and Code Administrators (BOCA model code Article 25), and ASTM (Subcommittee EO6.25) in the United States, the Department of Public Works of Canada, the Building Research Establishment in Great Britain, the French government regulations for the design and construction of public buildings (Moniteur Officiel, 1974), the International Council for Building Research Studies and Documentation (CIB/W60), the International Organization for Standardization (TC59/SC3), and consortia of property owners, developers, and product manufacturers (such as ORBIT-2 in North America).

They addressed questions of responsibility, procedure, and method, including how to organize the tasks and process for specifying the project requirements, identify essential performance requirements, measure achievement of appropriate overall performance, decide upon procedures and tools for measuring and levels of quality of measurement, and use the data for building evaluation and diagnosis of problems.

Characteristic of all these initiatives is a focus on how a facility will work for the people who own, occupy, or visit it: How will the building affect the health of its occupants? Will it be suitable for an organization with dense use of information technology, such as personal computers, word processors, large computers for managing data bases, and local area networks? Will the building fabric last long enough? Will it use energy efficiently? If it is remodelled and rehabilitated, will it be sufficiently safe for life and property?

## MANY CONSTITUENCIES FOR STANDARDS DEVELOPMENT

Each of these successively more comprehensive views of building performance was driven by a somewhat different constituency. The writing of standards for building products, materials, and furnishings was led by manufacturers, public interest groups, and building code officials. Standards for the performance of building systems, such as for ventilating, air-conditioning, energy conservation, and indoor air quality, were led by professional associations such as the American Society of Heating, Refrigeration, and Air Conditioning Engineers (ASHRAE) in the United States. The Human Factors Society and the Computer and Business Equipment Manufacturers Association have collaborated under the leadership of the American National Standards Institute (ANSI) in the development of a draft standard for the physical setting for using video display terminals.

## STANDARDS FOR RATING OR DIAGNOSING OVERALL PERFORMANCE

A new kind of "standard" is emerging to deal with the complexity of the overall performance of buildings, of calculating the economics of building construction, of environments with high technology, and of problems involving nuclear safety. Standard practices and standard guides have emerged that prescribe procedures and give guidance and references, and are unlike the rigid test and measurement procedures prescribed in standards for building materials and parts.

## RATING FOR OVERALL PERFORMANCE

Rating of overall performance is now being used for regulatory purposes. A form of rating procedure is used to determine whether the proposed adaptive reuse of a building will, overall, have life and property safety and security equivalent to those of a new building. This is now used in New York City Public Law 5, the BOCA model code Section 25, and the laws of at least eight U.S. states. When a reused building cannot be made to comply with codes in all respects, these laws and codes prescribe how to determine whether the level of safety and security, *overall,* is rated *equivalent* to that in a building that would comply in each separate respect with the requirements of the building code.

Rating is also being used for commercial purposes. The buyer of a $12,000 automobile typically has more comparative information about the purchase than the buyer of a $12 million office, because buildings aren't rated within price or style categories as are cars in *Consumer Reports.* The working drawings and specifications for a building give plenty of detailed facts about its parts but no direct basis for making comparative, overall judgments.

A method for assembling comparative information about buildings was developed by the ORBIT-2 project with respect to the impact of information technology on buildings and organizations. The ORBIT-2 project also developed guidelines for determining whether office facilities are likely to be suitable for high-technology users. Some large corporations are using these guidelines as de facto standards, enforced when these companies select which properties to rent or buy.

The ORBIT-2 approach first determined the relative importance of 17 key performance requirements, listed below. The first group of requirements have to do with *organizational issues*:

- *Change of total staff size.*
  The need to accommodate changes in total population of the organization/department/group.

- *Attraction or retention of highly qualified work force.*

- *Communication of hierarchy, status, and power.*
  The importance of having people recognize differences in rank, status, or power between people in the organization/department/group.

- *Relocation of staff.*
  The need to physically relocate the workplace of individual workers or small groups within the facility, without substantial inconvenience or cost.

- *Maximization of informal interactions.*
  The need for informal and spontaneous interaction and face-to-face communications among staff.

- *Human factors: ambient environment.*
  The need for adjustment of lighting, air quality, temperature, acoustics, etc.

- *Image to the outside.*
  The need to present an appropriate image to visitors from outside the organization.

- *Security to the outside.*
  The need to protect information and/or valuable objects from outsiders.

- *Security to inside.*
  The need to protect information, such as data in files or on computers, from insiders.

The remaining requirements have to do with information technology issues.

- *Connection of equipment.*
  The need to be able to connect all or most electronic workstations to networks, mainframes, or other electronic equipment not located at the workstation.

- *Change in location of cables.*
The need to be able to change easily the locations of the endpoints of cables that connect electronic equipment. (This applies both to endpoints at people's workstations and to endpoints in equipment rooms.)

- *Equipment that is environmentally demanding.*
The need to provide special environmental conditions such as special cooling, floor load capacity, humidity control, dust control, acoustics, freedom from vibration, etc., because of the presence of information technology equipment. (Provision of electrical power is not included in this requirement.)

- *Protection of hardware operations.*
The need to ensure that operations not be interrupted, even for a few seconds, because of failure of a computer or other information technology; and/or for data to be protected against loss, delay, change, or misrecording due to problems with computer or related hardware. (Software aspects of computer security are not included under the issue.)

- *Demand for power.*
The need for primary and secondary electrical power capacity and feed, including vertical and horizontal on-floor distribution.

- *Relocation of heat-producing equipment.*
The need to be able to relocate heat-producing information technology equipment frequently to unpredictable locations within the office work areas.

- *Human factors: workstations.*
The need to provide ergonomically appropriate workstations, with suitable furniture, equipment, and task illumination and sufficient horizontal and vertical space for all information technology equipment.

- *Telecommunications to or from outside.*
The need for large volumes of uninterrupted telecommunications to or from outside the building (including voice, data, or video, etc.)

The relative importance of each requirement varies from organization to organization, and any particular organization will likely change its own priorities over time.

In the ORBIT-2 rating process, each building or facility is rated for how well it can perform vis-à-vis each of these 17 performance requirements. ORBIT-2 provides, for each performance requirement, a range of typical combinations of physical or design features that, taken together, meet the requirement to some degree. Each combination of features is one that is typically found in existing facilities in the field. Each of these typical combinations of features has a performance score assigned to it for its effectiveness in meeting the performance requirement. Below are three entries from a sample listing:

*Protecting hardware operations*

| *ORBIT-2 rating* | *Actual rating* | *Combination of features* |
|:---:|:---:|:---|
| *9* | — | • *Uninterruptible power supply with adequate space and floor load capacity for expansion. Generator capacity or independent reserve source of power available on site, with capacity sufficient for full information technology operations. Secondary backup transformer. Buss power supply. Separate grounding for information technology equipment, direct to ground field.* |
| | | • *Either cables for information technology are in raceways shielded against induced current or there is adequate space to separate these cables from electric power supplies; this applies in both primary and secondary runs; generous clearance in ceiling space or access floor with more than 6 in clear is acceptable for this. Generous space available in distribution closets to service equipment. Highly effective antistatic floor covering at all workstations and wherever equipment for information technology is used. Effective humidity control in HVAC system.* |
| *7* | — | • *Uninterruptible power supply. Emergency generator capacity sufficient for the information technology equipment that must operate during emergency conditions. Buss power supply. Separate grounding for information technology equipment, direct to ground field.* |
| | | • *Some space available in distribution closets to service equipment. Shallow access floor, less than 4 in clear, or ceiling space available to run the separate clean power for each equipment item and other power and to separate all power lines from information technology cables. Good quality antistatic floor covering at all locations equipment for information technology is used. Effective humidity control in HVAC system.* |

5     —    • *Uninterruptible power supply. No emergency generator capacity. No special provision for grounding information technology equipment, but otherwise good electrical standards.*
- *Shortage of space in distribution closets. Space available only in ceiling and/or in floor ducts for running special power supplies. Cables for information technology must be induction shielded. Antistatic pads used at workstations with electronic equipment. No effective humidity control in HVAC system.*

ORBIT-2 also provides rules for adjusting the rating score to provide for minor differences between the combinations of features found in the facility and the closest combination of features rated in the guide. Below is an example of how these rules operate:

*The goal is to ensure that computer operations not be interrupted, even for a few seconds, because of failure of power supply, and that data or software be protected against loss, delay, change, or misrecording due to power surges or static electricity or induction in cables for information technology. This is critical in organizations where the ability to conduct business is dependent on electronic media (e.g., finance, insurance, banking) or where loss of records could be catastrophic.*

*Positive characteristics:*

Uninterruptible power supply.
*Backup battery system to prevent power loss even for an instant while switching over to alternative power source such as backup generator. Typically do not provide power for more than 1 to 2 hours.*

Backup generator or alternative source of power.
*Provide alternative sources of power that, in the event of failure of the supply from the electric power utility, enable equipment and building systems to operate with only a few seconds or minutes of downtime. May be provided in conjunction with uninterruptible power supply, which complements but does not replace backup generator or alternate source of power.*

Backup transformer.
*Provides an alternative in the event of failure of the main transformer from the utility. If a transformer burns out it may take weeks to get a replacement, and during that time all or a portion of the facility would be without power if there were no backup transformer available.*

*Instruction for adjusting the ORBIT-2 rating.*
If the building has any of the above characteristics, but the combination of features you are using—i.e., the one closest to the situation in the building you are rating—does not, then raise the ORBIT-2 rating by one point. If the combination of features you are using has any of the above characteristics, but the building does not, lower the ORBIT-2 rating by one point.

ORBIT-2 also provides other adjustments, including some of half a point each, and it provides additional typical combinations of features, each with an ORBIT-2 performance score. Multiplying a performance score for each requirement times its relative importance gives a weighted performance score. The procedure for calculating the rating is consistent with the practice now being balloted within ASTM.*

## ASTM SUBCOMMITTEE EO6.25 ON OVERALL PERFORMANCE OF BUILDINGS

Both the BOCA code for rehabilitating buildings and the ORBIT-2 project take an approach to rating performance that is fully compatible with the family of standard practices and guides being developed by ASTM Subcommittee EO6.25 on Overall Performance of Buildings. These provide a *framework* for tests and measures to evaluate the overall performance, for specified purposes, of an existing building or of a facility within a building. They distinguish between two kinds of measures of performance.

1. The *rating* of a building or facility to determine how it meets specific requirements for overall performance vis-à-vis specific objectives such as the following:

---

* The ORBIT-2 procedure, including the grouping of key issues and combinations of features, has been further developed and revised, drawing on experience in the field. The version being balloted within ASTM incorporates these developments, and is consistent in principle but different in details from the text above.

- Aesthetics, image, and character of public spaces
- Building operation and maintenance
- Condition of historic fabric
- Energy use, conservation, and reliability
- Functional effectiveness of facilities of specific hospital departments
- Functional effectiveness of office buildings for various types of tenants
- Health, well-being, and comfort of office occupants
- Space utilization in hospitals
- Space utilization in libraries
- Suitability of offices for organizations that are changing rapidly
- Suitability of offices for organizations that have much new information technology
- Suitability of museums for various types of traveling exhibits

2. The *in-depth evaluation* of specific aspects of the overall performance of a building or facility

## IN-DEPTH EVALUATION

In-depth evaluation, such as that required when significant problems must be diagnosed or when the success of various design decisions must be assessed, has been the subject of development work for several decades. A landmark project was the GSA/PBS Building Systems Program, which has been described and assessed in two documents by Francis Ventre, published in Washington, D.C., in 1983 by the National Bureau of Standards (NBSIR 83-2262 and NGSIR 83-2777). Work by the Architectural and Building Sciences Division of Public Works Canada, under the direction of Peter Mill, has been described in a paper by Volker Hartkopf, Vivian Loftness, and Peter Mill titled, "The Concept of Total Building Performance and Building Diagnostics," in ASTM Special Technical Publication 901: *Overall Building Performance: Function, Preservation, and Rehabilitation.*

## IMPACT OF INFORMATION TECHNOLOGY

By the end of the 1980s, a majority of office workers are expected to have some form of video display terminal at their workstation; some say nearly all will have some form of visual display unit by the mid-1990s. Already today, many organizations have dense installations of information technology, such as personal computers, terminals, printers, modems, telecopiers, photocopiers, digital telephones, and supermicrocomputers, and new equipment is added frequently.

These organizations are different from what they were before technology was introduced and different also from organizations that have not yet acquired much information technology. For instance, compared with how they functioned before they introduced information technology, groups with dense installations typically have more frequent changes in how they are organized and in how they work, hold more meetings, relocate people in the building more frequently, and have fewer levels separating the top decision makers from the junior staff.

The buildings they need are also different. These organizations have a whole range of new requirements, involving such matters as building shape, type and location of vertical circulation, access to views, partition and screen systems, furniture, lighting, and air quality. In addition, their buildings must accommodate the hardware itself, with its cables, heat, and pollutants. De facto standards for many of these aspects of building performance are now emerging.

## EXPECT MANY STANDARDS FOR OVERALL PERFORMANCE

Within the next few years there is likely to be a proliferation of standards for rating or in-depth evaluation of specific aspects of the overall performance of buildings and the facilities in them. These will stimulate the development and marketing of low-cost, portable equipment for conducting key measurements at the building site, such as for levels of $CO_2$ and for nonionizing electromagnetic radiation. Organizations that occupy large amounts of space and that use high-technology equipment, such as multinational corporations and governments, will use their economic power to enforce their internal standards on landlords and property owners, so creating de facto standards. Concurrently, the consensus-based standards from ASTM, and other standards developed through ANSI, ASHRAE, CSGB, and ISO, will provide a publicly available framework for these standards and guides for the small company and small property owner.

CHAPTER ELEVEN

# OVERVIEW AND ASSESSMENT OF THE BUILDING PROCUREMENT OPTIONS IN NORTH AMERICA FOR HIGH-TECHNOLOGY COMPANIES

*COLIN H. DAVIDSON*

**R**ather than enumerate in checklist form the options that are available when high-tech facilities are to be procured, this chapter describes the nature of procurement decisions and presents some of the principles underlying them: their scope, their influence, and their timing. Careful procurement is shown to have an impact on the quality and the cost of the facilities being acquired—particularly if the high technology to be sheltered in these facilities introduces unusual programmatic or environmental requirements.

The context for decision making by the building owner is presented first,[1] by a description of pertinent aspects of the traditional building industry. Then the subject of procurement itself is broached, in general terms, in order to shed light on the broad range of options available. A glance at the "high-tech" aspect of high-tech facilities then shows that they possess two characteristics that are likely to require special consideration when procurement decisions are made; at this point it becomes possible to assess suitable procurement approaches. By way of conclusion, the importance of planning the procurement decisions is stressed, showing that as the building owners make these decisions, they are in fact doing no more than exercising fully their special role in the building industry.

## THE TRADITIONAL BUILDING INDUSTRY

The building industry *as a whole* is described in management terms as a "multi-industry" because it is composed of a number of individual professional agencies and industrial enterprises, each existing for its own reasons and, hopefully, operating over a long period of time, as it carries out repetitively the many building-related tasks for which it possesses the necessary skills. At the center of this multiindustry are the block of architectural and engineering offices (operating on a declared-fee basis) and also the block of contracting, subcontracting, and manufacturing companies (operating on a concealed profit-within-prices basis). Around them, other enterprises, such as testing and controlling laboratories, utility companies, and mortgage and banking corporations are clustered. They all share the common feature of surviving in the long term while participating in individual projects in the short term; they further their own particular objectives while participating in, and thus accepting—to a degree at least—the imposed objectives of each of the projects (imposed, that is, by the building owner's procurement decisions).

Looked at from the point of view of the *individual projects,* the building industry gives rise to "temporary multi-organizations"—task-oriented regroupings of enterprises brought together into a system of contracts and dependencies by the intending building owner(s) (Figure 11-1). By their very nature, these organizations represent *discontinuous* and *dispersed* "teams," in which the participating firms' objectives are often in more or less overt conflict with the project's objectives. That successful building projects *do* occur at all, and that firms *do* survive in their part of the building marketplace, are due to the fact that all participants in the traditional building process know what is generally expected of them,[2] thus compensating for the seemingly impossible mix of interdependency and uncertainty[3] that is inherent in the temporary multi-organizational structure of the so-called building team.

## TRADITIONAL PROCUREMENT

The intending building owners must purchase the facilities they require from this building industry; as best they can, they must stay within the traditional bounds of what the industry can produce (even if they do not un-

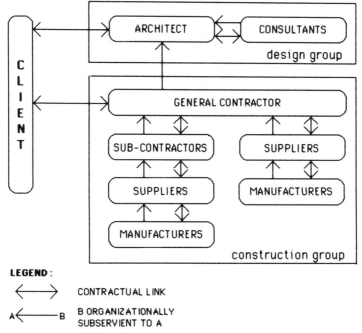

LEGEND:

←——→  CONTRACTUAL LINK

A←——B   B ORGANIZATIONALLY
        SUBSERVIENT TO A

**Figure 11-1.** Organigram of the traditional *temporary multi-organization,* or so-called building team, brought together to carry out a traditionally commissioned building contract. Note (1) that this grouping is not formed at one moment but is gradually constituted as and when participants are required to intervene, and then progressively dissolves as their participation is finished—it is therefore "discontinuous," (2) that only the links shown actually exist formally and other plausible ones in reality do not occur—this grouping is also, therefore, "dispersed" (source: *Encyclopedia of Architecture: Design, Engineering and Construction,* Joseph A. Wilkes and Robert T. Packard, eds., copyright © 1988, John Wiley & Sons, Inc.).

derstand these bounds very well), and they must organize the participants they wish to work with into compatible groups. In other words, they must *design* the temporary multi-organization they need—and design them by their procurement decisions (Figure 11-2).

The first of the procurement decisions is whether to purchase or to lease the required facilities—a decision that is based on a study of past experience and on in-house objectives and constraints. The second set of decisions, which is imperative if the purchase option is chosen (whether deliberately or by default—because, for example, no suitable facilities are available for lease), concerns whether the facilities are purchased off the shelf or constructed to designs. The very existence of these decisions reflects a particular characteristic of new buildings—namely, that they are purchased before they are built. It is also a set of decisions that, in effect, corresponds to the degree of initiative the owners want to take regarding the specificity of the facilities and concerning their involvement in the procedures required for obtaining them.

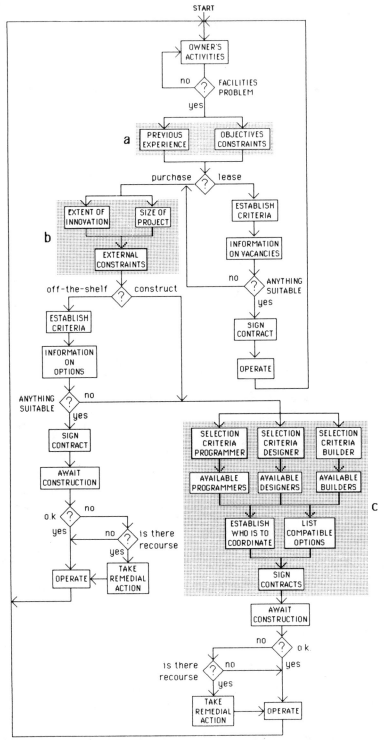

**Figure 11-2.** Procurement decisions in the facilities acquisition process. Note that preparatory work takes place in three blocks, in readiness for (a) the purchase or lease decision, (b) the off-the-shelf or construct-to-designs decision, and (c) the design and construct phase.

The third set of decisions determines the organizational base for the process that is to be set in motion if purpose-designed and purpose-built facilities are preferred. The decisions bear on functional programming (listing of requirements to be satisfied by the intended facility), architectural and engineering design (how the design task is to be carried out, and in what relation to the other functions), and construction (what the responsibilities for construction are to be and how they are to be shared). Figure 11-3 (a through i) shows some of the organizational changes that can be introduced into the temporary multi-organization as a result of variations in this third set of procurement decisions. It should be noted that combinations of the variants shown in Figure 11-3 can be envisaged; for example a with b or f, b with f, etc. It should also be noted that variants a, b, f, h, and i allow owners to increase their role in, and therefore control of, the building process, whereas c reduces it (even though d and e reestablish some degree of owner control); g places the initiative with a contractor/manufacturer consortium—which is an advantage to the owners only if their facilities requirements can be satisfied by the offerings of the consor-

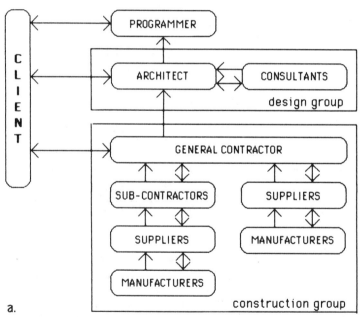

**Figure 11-3.** Organigrams of variants to the temporary multi-organization resulting from procurement decisions by the building owner (source: *Encyclopedia of Architecture: Design, Engineering and Construction,* Joseph A. Wilkes and Robert T. Packard, eds., copyright © 1988, John Wiley & Sons, Inc.). These should be compared with the traditional organization presented in Figure 11-1; they follow from decisions of block (c) in Figure 11-2.

(a) *Programming approach*—differs from the traditional organization by the addition of a programming consultant outside the owner's organization, with responsibility for listing the owner's functional requirements in terms that can be used by the facilities' designers.

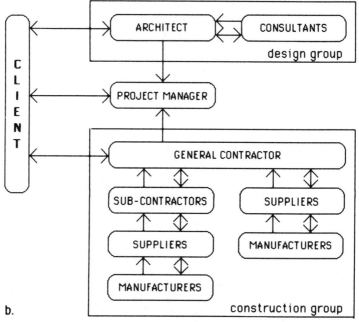

b.

(b) *Project management approach*—adds a project manager to coordinate design and construction tasks for the owner, without imposing this responsibility on the architect or the contractor.

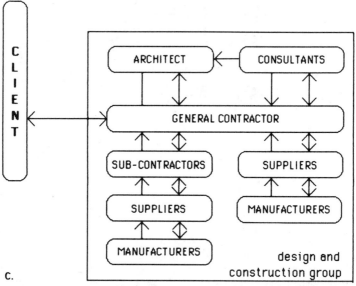

c.

(c) *Design/Build organization*—places full responsibility for design and construction with a construction company or developer/builder, who receives instructions about requirements directly from the owner together with information about controls and evaluations that will be imposed prior to acceptance.

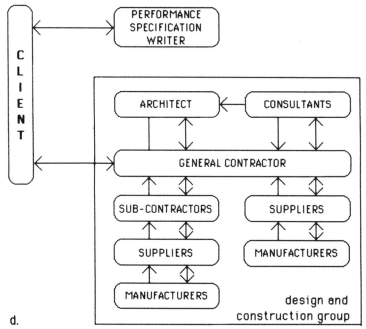

d.

(d) *Developer proposal approach*—adds a performance specification consultant to the design/build organization so that the owners' intentions can be described better and their satisfaction properly monitored.

tium sponsors, and only if they obtain good time and cost conditions out of their procurement negotiations.

In summary, the third set of decisions allocates responsibility for design, construction, administration, and financing and determines the basis for the commitments required later for the actual construction (see Figures 11-5 and 11-6).

The second set of procurement decisions, as was mentioned previously, concerns the choice between the purchase of off-the-shelf facilities (pre-designed and/or prebuilt) and the purchase of facilities constructed-to-designs. The significance of this choice lies in the fact that, apart from light industrial or commercial facilities, small office buildings, or tract houses, virtually all building is "bespoke." Indeed, it is a characteristic of the building part of the building industry that it sells services and not products; put another way, the intending building owner must make the building purchase commitment *before* the product exists, before its quality and appropriateness can really be known—even if its price is fixed in advance by contract. Obviously, if only for this reason, it is important for the purchasing activities—the *procurement*—to be conducted with foresight! Figure 11-4 illustrates various initiatives by the building owner, and responses by the building industry, in cases of off-the-shelf and design-and-build procurement.

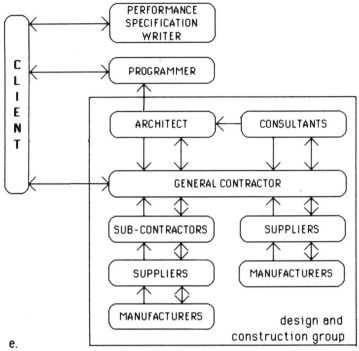

e.

(e) *Design/build bid approach*—adds a programming consultant to the developer proposal approach, with a double mandate to prepare a functional program and to establish fair rules for competition between construction companies' bid proposals.

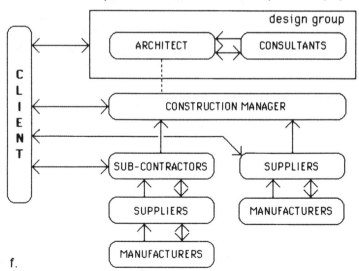

f.

(f) *Construction management approach*—replaces the general contractor by a management consultant retained by the owner as agent; if this consultant does more than coordinate the activities associated with construction to include design coordination, then the role merges with that of the project manager (variant b above).

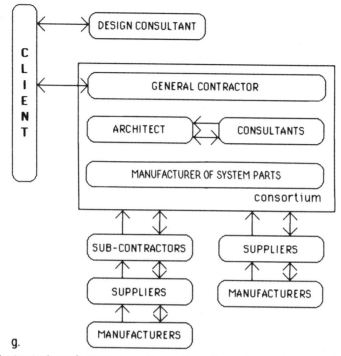

g.

(g) *Contractor/manufacturer consortium*—allows the contractor to develop new forms of building with a manufacturer, which are then proposed to owners (who may turn to design consultants for evaluation advice); this is the case of preengineered buildings or contractor/manufacturer-sponsored prefabricated building methods.

## HIGH-TECH FACILITIES

It has often been said that as long as the building industry is required to perform traditionally (that is to say, produce traditional buildings under traditional constraints of quality, time, and cost), the inherent checks and balances within the industry on the one hand, and its accumulated know-how on the other, enable it to meet the needs of its clients (the building owners)—*provided the latter make predictable procurement decisions.* Before asking whether such predictable decisions are still possible with high-tech facilities, it is necessary to see what their particular "high-tech" characteristics are (if any), and how these facilities are distinguished from other facilities built to satisfy more routine requirements.

Appearances are deceptive, and from the point of view of procurement, high-tech *styling* must be ruled out immediately as a distinguishing characteristic. Since the beginnings of the Modern Movement, architects have been designing "modern-tech" buildings, albeit with more or less success (the less successful cases usually leading to premature and costly maintenance); in some cases, great efforts were (and are being) made to bend the

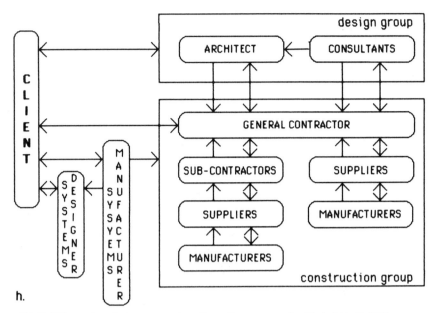

(h) *Building subsystems approach*—allows the owner, who likely has multiple long-term facilities requirements, to have special building subsystems designed and manufactured for specification and use in individual building projects (this approach has variants depending on whether the subsystems can be nominated or must be competed for).

still-traditional technology of the building industry to produce buildings that *look* machine-made and high-tech.[4]

High-tech facilities may, however, have either or both of two characteristics that stem directly from the functions to be housed and that predetermine the context for all related procurement decisions. The first of these characteristics—which can be called "hard"—derives from the *physical* nature of the high-technology operations for which the facilities are being built. For example, special environmental controls may be needed to allow microscopic manufacturing processes to be undertaken, or illuminating installations may have to be flexible to avoid dazzle at subsequent rearrangements of graphics cathode ray tube workstations; access to special gases and fluids may be needed, or high-capacity data communications networks may have to be installed throughout.

The second of these characteristics—called "soft"—derives from the *organizational* nature of the high-technology operations. For example, is there a requirement regarding the timing of facilities' completion—for example, that it coincides with the construction (by others) of independent conference or recreational facilities? Or should the long-term ownership of the facilities guarantee access to shared management services that

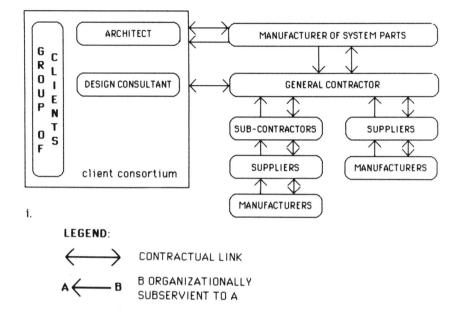

i.

LEGEND:

←——→  CONTRACTUAL LINK

A ←—— B   B ORGANIZATIONALLY
SUBSERVIENT TO A

(i) *Organization of clients' consortium for system building*—associates the architectural and systems design functions (found in the building subsystems approach) more closely with the owners, who are here grouped into a consortium, presumably because they share multiple long-term facilities requirements (this approach also has variants, depending on whether the systems can be nominated or must be competed for).

were planned from the outset and that are reflected in the layout of the various spaces?

## HIGH-TECH PROCUREMENT

Procurement decisions enable the building owners to form temporary multi-organizations in a way that ensures them the best chance of obtaining value for money by grouping designing, manufacturing, and building skills in a hierarchy of authority, and vested interests in a network of subordination most favorable for the project.

When high-tech facilities are to be built to meet special hard requirements, the chosen procurement strategy must give a prominent role to the engineering skills associated with the needed support technology. The interests of the specialist contractors who are to carry out the corresponding work (and whose role is critical in ensuring its quality) must also be pro-

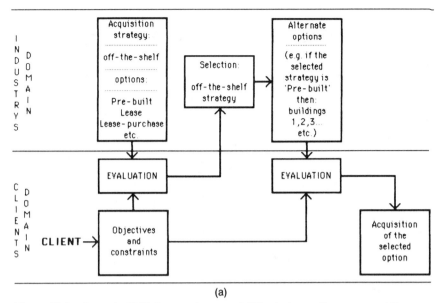

(a)

**Figure 11-4.** Spread of initiatives and responsibilities between the owner and the building industry, in the procurement of (a) off-the-shelf facilities or (b) constructed-to-designs facilities.

tected, otherwise they may be induced to perform their work shoddily (because, for example, they sense that their profit margins are being eroded by the impacts of the other building participants on the operations for which they are responsible contractually—impacts, it should be remembered, that are largely controlled by the adopted procurement strategy). Incidentally, when there are very specialized hard requirements, it is unlikely that suitable facilities can be leased or purchased off the shelf; purpose-designed construction arrangements must almost inevitably be made, where programming, design, and construction options are organized around satisfaction of these hard requirements by the building owners, through their procurement decisions.

When special soft requirements exist, the nature of the procurement decisions changes. Instead of the building owners' seeking only to set up building teams best geared to meet their particular needs, they must now coordinate their own procurement decisions with those of other building owners as well. They must ensure that a set of facilities, whose completions must be coordinated if any one of them is to operate beneficially (as, for example, in the case of campus-like science parks), are procured compatibly; or they must ensure that postoccupancy ownership arrangements (which are also part of the set of strategic purchasing decisions) preserve an appropriate level of interest in the coordinated management service functions originally planned for (as, for example, in the case of industrial incubators).

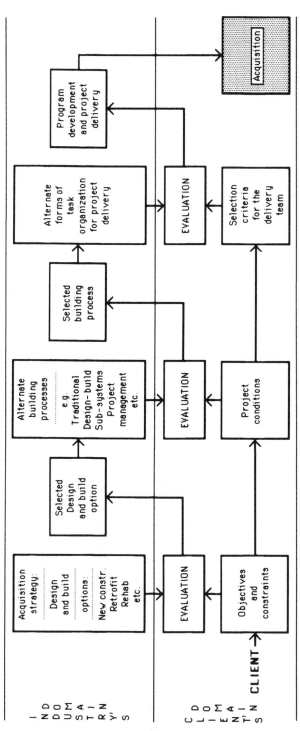

INDUSTRY'S DOMAIN

CLIENT'S DOMAIN

CLIENT

Acquisition strategy:
Design and build options:
New constr.
Retrofit
Rehab
etc.

→ EVALUATION ←

Objectives and constraints

Selected Design and build option

Alternate building processes
e.g.
Traditional
Design-build
Sub-systems
Project management
etc.

→ EVALUATION ←

Project conditions

Selected building process

Alternate forms of task organization for project delivery

→ EVALUATION ←

Selection criteria for the delivery team

Program development and project delivery

Acquisition

Coordinated decisions are not typical of building owners, who are usually unaware of (and able to operate effectively *without even having to be aware of*) decisions being made around or after them. Indeed, the multi-industry/temporary multi-organizational structure of the building industry has grown up over the ages as a response to this "go-it-alone" behavior on the part of its clients.

Some kind of consortium approach by the concerned building owners may provide an adequate answer to the coordinated procurement requirement of the science park (as shown in Figure 11-3i, with or without the system-building connotations). Integration of responsible short-term construction ownership with long-term operational ownership of the facility would seem best in the case of the special requirements of an incubator (ruling out the delegated responsibilities of approaches c and g in Figure 11-3).

## CONCLUSION

Procurement decisions allocate responsibility for design, construction, administration, and financing between the participating members of the temporary multi-organization (Figure 11-5) and determine the basis for the corresponding contractual arrangements (Figure 11-6). It is these decisions that establish the power of the building owners to define their needs fully, and to ensure, by allowing them adequate recourse (see Figure 11-2), that the buildings they obtain do in fact satisfy their needs. In the case of high-tech facilities, the risks of error by the building industry are greater, and the need for the building owners' self-protection is higher than with "ordinary" buildings; in all cases, however, wise owners reserve time and resources for their procurement decisions, *which they plan as soon as they have their first thoughts about a new facility.* Procurement cannot be left to chance, nor can it be left to later on.

## ACKNOWLEDGMENTS

The author is indebted to Dr. Rashid Mohsini and Mr. Mario Gagné, with whom he had many occasions to discuss the topic of procurement. Figures 11-1, 11-3, and 11-4 are based on an unpublished report by Dr. Mohsini, prepared in collaboration with the author; Figure 11-2 is based on a working document prepared by Mr. Gagné. Figures 11-5 and 11-6 are inspired by work of Prof. David Haviland (see Bibliography).

## NOTES

1. In French, interestingly, a building owner is called the "maître d'ouvrage" (the master of the work) in contrast to the designer, who is the "maître d'oeuvre" (master of the works); procurement is referred

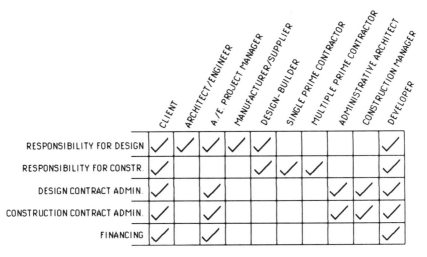

**Figure 11-5.** Procurement options: possible allocations of responsibilities for design, construction, administration, and financing between participating organizations.

| BASIS FOR CONSTRUCTION COMMITMENT | Complete construction documents<br>Scope construction documents<br>Outline construction documents<br>Performance documents |
|---|---|
| BASIS FOR PRODUCTS ACQUISITION | Conventional documents<br>Subsystems<br>Precoordinated subsystems<br>Building systems |
| BASIS FOR CONTRACTS AWARD | Negotiated<br>Competitive, one-step<br>Competitive, two-step |
| BASIS FOR COMPENSATION | Lump sum<br>Cost-plus-fee<br>Guaranteed max. price<br>Other forms |

**Figure 11-6.** Procurement options: possible bases for contractual arrangements.

to as "maîtrise d'ouvrage" in the implied sense of "mastering the tasks of the building owner."

2. "By traditional building is implied the bases of design, organization and execution of building which have come to be recognized as normal practice over a considerable period in any country or region. It is usually characterized by the fact that all operations follow a set pattern known to all participants in the actual building operation, and by dependence on skilled craftsmanship for interpretation of instructions

and execution of work"—definition given in *Government Policies and the Cost of Building* (Geneva: UN/ECE, 1959).
3. See, for example, *Interdependence and Uncertainty, a Study of the British Building Industry* (London: Tavistock Publications, 1966).
4. Paradoxically, the sponsors of the machine-made "industrialized" buildings of the post–World War II years carefully dressed up their products with traditional styling.

## BIBLIOGRAPHY

Davidson, Colin H. "Importance of Conventions in the Industrialization of Building," in *Sistematizacion y Industrializacion de la Vivienda,* Mexico (D.F.), Universidad Autonoma Metropolitana, Xochimilco, 1976.
Davidson, Colin H. "The Building Team," in *Encyclopedia of Architecture: Design, Engineering and Construction,* edited by Joseph A. Wilkes, and Robert T. Packard. New York: John Wiley and Sons, 1988, pp. 509–515.
Davidson, Colin H. and Rashid Mohsini. "Building Procurement—a Strategic Organization and Management Decision." In *Managing Construction Worldwide,* vol. 1: *Systems for Managing Construction,* edited by Peter R. Lansley and Peter A. Harlow. London and New York: E. & F. N. Spon, 1987, pp 28–39.
Glover, Michael, ed. "Building Procurement; Proceedings of a Workshop," *IF Occasional Paper Number One,* published jointly in Montreal and Champaign, Ill., by the IF Team (University of Montreal) and the Construction Engineering Research Laboratory, 1974.
———. "Alternative Processes; Building Procurement, Design and Construction," *IF Occasional Paper Number Two,* published jointly in Montreal and Champaign, Ill., by the IF Team (University of Montreal) and the University of Illinois, 1976.
Haviland, David S. *A Critique and Commentary on Procurement Procedures for Systems Construction,* unpublished report. Troy, N.Y., Rensselaer Polytechnic Institute, Department of Architecture, 1981.
Mohsini, Rashid. *Building Procurement Process, a Study of Temporary Multi-Organizations,* unpublished report. Montreal, University of Montreal, Faculté de l'Aménagement, 1985.
Norsa, Aldo, ed. "Answers for the Building Community: Optimizing the Choices," *IF Occasional Paper Number Three,* published jointly in Montreal and Champaign, Ill., by the IF Team (University of Montreal) and the University of Illinois, 1976.
Roberts, C.J.B. *Project Analysis and Organization Design in Building: An Investigation Into the Performance of Building Projects,* unpublished report. St. Louis, Washington University, 1972.

CHAPTER TWELVE

# THE GENERAL SERVICES ADMINISTRATION'S ADVANCED-TECHNOLOGY BUILDINGS PROGRAM: A STATEMENT OF DIRECTION, BUILDING FEATURES, AND IMPLICATIONS

*DAVID B. EAKIN*

## BACKGROUND

In 1983 the General Services Administration (GSA), through the Public Buildings Service (PBS), committed to the development of an Advanced-Technology (Advanced-Tech) Buildings Program. This initiative emphasizes cost-effective innovation in all facets of building design, providing enhanced building performance capabilities for the total work environment. This encompasses traditional issues of building system efficiency, low maintenance, building security, and fire safety. It also focuses on occupant productivity by supporting concepts of office automation, advanced communications, space flexibility, and improved reliability of services. This is being done, not to be in compliance with a legislative requirement, but in response to our perceived mandate to assure that federal work space represents the highest levels of efficiency and performance for its tenants, the employees of the U.S. government. This mandate relates not only to new construction but to the quarter of a billion square feet of existing space within the PBS inventory.

The emphasis on improved building performance can be seen as evolutionary. Prior to the energy-conscious designs of the early 1970s most buildings were constructed emphasizing limited functional interests, influenced by basic performance criteria. In architecture, concerns over appearance and space assignment usually governed project direction. Form ruled over function. For engineering, the issues centered around assuring adequate capacity and comfort control. No attempt really existed to involve other functional issues related to the work environment. It was also the case that most design disciplines worked independently, with the only evidence of coordination being associated with the traditional assurances of structural support, space assignment, clearances, and the exchange of information concerning thermal loadings and power needs.

With the energy crisis of the 1970s, there came into performance criteria, perhaps for the first time, a comprehensive address of a single functional goal by all disciplines—that of energy conservation. Architectural envelopes became better built with the use of better sealing techniques, more insulation, and imaginative glazing treatments for passive solar benefit. Mechanical systems became more efficient with improved control. And electrical systems were designed for reduced lighting and power consumption. But as in years past, advancing the building's support of the interior work environment was really not addressed. This was left to the tenant: enter the gray metal desks, mismatched finishes, hot spots, wire-tripping hazards, and a sense of disorganization.

The designer's disassociation from the working environment also led to building systems that were sufficiently inflexible to compromise building efficiency and performance when used outside the original design assumptions. For much of the building industry today, not much has changed. In fact, there are buildings being completed today that offer only a few functional enhancements over those provided our great-grandparents.

## GOALS AND OBJECTIVES

With the advent of the microcomputer came an opportunity to change the way people do office work, and in turn, how buildings must be designed. Now there are new demands on building features and environmental services. More than ever, designers should be influenced by how tenants do business. And in most cases, they must be even more concerned with how tenants might do business in the future.

## Cost-effectiveness

The Advanced-Tech approach to the work environment is different also from previous construction efforts in its sensitivity to the costs of doing business. Fifteen years ago, most design emphasis was directed to providing a building within budget, offering a required amount of generic space. Compared with the total cost of operating a service-oriented business (or organization), this represents only 3 to 4% of the monies that will be committed over a 25-year life. With the energy crisis, our cost sensitivity was increased to include building operation. This led to the use of life cycle costing as a means of selecting building systems and components. Still this new operational focus could only increase our business cost perspective by 5%. What has been missed until now is the recognition that an information-based organization has an overwhelming 91% of its expenses in the salaries of its employees. It is here that we must direct our design attention to try to improve performance and efficiency. The potential economies of occupant productivity are great. For if by committing a 10% additional cost to the building's construction we can save 1% in employee efficiency, the return on our investment is almost 3 to 1.

## Productivity

Annualized costs of an information-oriented business can also be put into perspective by recognizing the overwhelming costs of occupant salaries and associated productivity. In relation to a building's square footage, today's combined costs of utilities, equipment maintenance, building protection, and other building operations may be between $4 and $6/sq ft. The annual cost for occupant salaries is typically between $200 and $500/sq ft. It can even be shown that a 1- or 2-day disruption to an occupant's function can have a cost impact equal to the occupant's prorated share of the building's annual electricity costs. Clearly, the occupant must be offered an efficient, comfortable environment with assurance of reliable services.

## Flexibility

Perhaps the single biggest functional need relates to flexibility. Since most government buildings have a life expectancy between 40 and 100 years, it is important to make them as adaptable as possible to changing technologies. In the electronics industries, for instance, technology changes every 3 to 5 years. Hence, from the time it takes to plan for a new building to the time it is occupied—some 4 to 6 years—office automation technology may have changed twice. In the building's expected life, there may be dozens of equipment generations. The design challenge is to provide building space and systems that are compatible with near-term tenant requirements, while being flexible enough, in space and system capacities to support technologies of the future.

## Reliability

Assured availability of services is also becoming more critical. Since the working medium of the Advanced-Tech office is the computer, power and environmental control will have to support computer operating needs as a critical function. The more eggs that are put into the office automation basket, the more critical will be the role of supporting systems.

## Energy Conservation

The functional goal of achieving an energy-conscious design is very much a part of the Advanced-Tech building. The objective is to provide innovative concepts, addressing higher efficiency equipment and the use of heat-recovery systems. Each building system is given individual energy targets to ensure achievement of energy performance. Overall building annual energy performance goals are not recommended because of technical and administrative problems of enforcement and because of the difficulty in applying them to retrofit construction. With the Advanced-Tech features of increased monitoring and centralized control of operating systems, energy performance should also be easier to manage and improve through building operations.

## Water Conservation

It has been speculated that water usage will be this country's next major crisis. In certain parts of the country, such as California, it is already a major issue in design. Our buildings must be developed to minimize water use through avoidance of need, limitation of waste, and possible reuse.

## Low Maintenance

Reduced operating and servicing costs are goals that can usually be achieved through design selection of durable materials and rugged equipment of few moving parts. Materials and protective treatments can be selected to minimize cleaning and abuse-related repair costs. Equipment and building systems can also be chosen for their history of proven low maintenance. However, the increasing use of sophisticated equipment operations can limit the overall savings potential and often requires specialized training. For such complex building systems, the use of multiyear service contracting should be considered. Increased use of service contracting may also help to ensure accountability of system performance and to offer improved reliability.

## Fire Safety and Security

Safety and protection are traditional goals of any building program, but the Advanced-Tech program is characterized by the broader application of more stringent standards and solutions, providing higher levels of safety and reliability.

## Scope Overview

The total work environment encompasses those conditions and functional interests that influence the occupant's work day. In addition to the superstructure and basic enclosure of the work space, special attention should also be given to issues of air quality and temperature control; ergonomically appropriate furnishings; availability and reliability of utilities, services, and communications; acoustic and personal privacy; security from hazards and abuse; and the office equipment resources to perform work efficiently. The new Advanced-Tech building then represents a broad base of functional issues covering all design disciplines. Wherever practical and appropriate, the most advanced available technology will be applied, not only to offer new levels of performance but to correct longstanding problems of basic service. This program offers PBS an opportunity to gain experience with the latest systems and devices that have great promise but whose application may not yet be common. Although there are potential risks, the possible benefits are great.

## PROGRAM ELEMENTS

To accomplish these objectives, PBS has established a multifaceted design and construction program. In summary, the program involves major efforts in research, prototype construction projects, criteria and policy development, and new design and construction practices. Through applied research findings and our own experience with state-of-the-art building features, the management/information loop is opened through design programming and criteria development to establish scope and project direction. It is then closed with post-occupancy evaluations to see if what we set out to do was accomplished and to verify that what we wanted to do was what we should have done. With adjustments to design and construction criteria, we can then fine-tune our approach and expectations on future projects. PBS is now in an evaluation phase of program accomplishments to establish those lessons learned that can be applied to future projects.

## Design Programming

To put form and substance to the term *Advanced-Tech* as applied to major construction projects, we developed a compilation of our "corporate knowledge" of what must, should, and could be done in the name of innovation. This was done in support of a process called "design programming." Design programming is critical to a project as it represents the detailed planning and direction phase, which is turned into a document, serving as the technical scope of work for the design architect/engineer. To form a generic base of programming direction, PBS not only pulled from its own in-house resources but acquired input from the National Institute of Standards and Technology (NIST). This has been further supplemented by design programming contractors and consultants charged to develop specific project direction on a series of prototype Advanced-Tech buildings.

In effect, the programming efforts have been centered around a matrix relating building systems to functional performance issues. The matrix relates each intercept of building system and function with the objective of directing the most innovative and advanced design solution. What this offers is the awareness that each building system must support all function requirements of the project. Or, said another way, each functional issue must now be supported by all building systems and features. Of course, some intercepts are without meaning, and some innovative solutions may be in conflict with other functional interests. The analogy of a crossword puzzle can be used, where the objective is to get all functional interests satisfied, without offering significant conflict and compromise to the overall solution.

## Prototype Construction Projects

To apply our programming concepts, we have identified a series of Advanced-Tech prototype buildings and made efforts to enhance design work already started. The Portland East Federal Building offered an opportunity to improve an already advanced building design with innovative building automation and telecommunication concepts. The combined effect represents "leading edge" technology that will offer the most functionally advanced building in our inventory. There was also a series of pending new office buildings that offered an opportunity to develop prototypical Advanced-Tech solutions on a more comprehensive basis: in Overland, Missouri; Long Beach, California; Oakland, California and Los Angeles, California. And finally, the Terminal Annex project, in Dallas,

Texas, and the Mart Building in St. Louis, Missouri, offered retrofit efforts that gave an understanding of what can be done to existing buildings. Through these projects, the diversity of tenant agencies, climatological influences, and site conditions offer a basis to quantify present limits and opportunities of Advanced-Tech construction. But these will not be the end of Advanced-Tech projects, as all future buildings will be designed for innovative building solutions. As technology changes are perpetual, so must be our commitment to apply and learn from new design concepts.

Even though it would be interesting to develop buildings based upon developmental concepts, this is not practical. The concepts that are being applied to our "prototype" buildings represent building features that offer a high probability of success. And, we can build only what we know how to build, with the materials and technology available to us today. Hence, most of the concepts represent recognizable state-of-the-art materials and systems. The difference comes in their application and in the integration of all functional objectives.

## Research

In order to improve our understanding of building sciences and to help advance developing concepts, PBS has initiated applied research efforts. Agreements have been established with the National Institute of Standards and Technology to support a wide range of research topics. The subjects have addressed innovative building features, office automation design, thermal integrity, environmental diagnostics, and base isolation structural systems. Although these studies have satisfied many program interests, further analysis is required for a growing list of technologies and problems. Major emphasis is required to establish cause and effect relationships in occupant productivity. There are also technology transfer issues of improving construction inspection techniques to assure that enhanced performances are realized.

## Design Criteria and Guidelines

In order to support the technology transfer needs of our agency, PBS is in the process of updating a Design Management Series of handbooks. Perhaps the most critical is the document "Quality Standards for Design and Construction" (PBS P 3430.1), which presents basic design policy and criteria. A second handbook of importance will be "Project Development for Federal Buildings" which will address the development of design directives for specific project applications. With these in place, PBS will be able to represent the lessons learned from our prototype projects. These references will also serve as the basis of addressing the evolving technologies of the future.

## Training

Establishing and maintaining expertise within PBS will be a continuing task to keep up with technologies and experiences. Of particular concern are the subjects of office automation, building automation, integrated telecommunications, and building diagnostics. There is also the necessity to ensure that PBS operating staffs have the knowledge to run and administer to the needs of automated systems. The use of long-term servicing contracts should allow for the necessary on-the-job training by service staffs, from the system manufacturer. Without a training commitment, Advanced-Tech objectives will not be realized.

## BUILDING FEATURES

It should now be apparent that the PBS Advanced-Tech building (sometimes called a High-Tech building) is not the same thing as is "a smart building," "an Intelligent Building," or "the office of the future." For these three terms relate primarily to the integration of building telecommunications. PBS Advanced-Tech buildings go beyond this narrow focus by considering the total integration of innovative technologies for all building systems.

The development of specific building features is a complex task, involving many program mandates and objectives, some of which can even be in conflict if not properly developed. In some instances, a measure of performance compromise is accepted so as not to defeat the service needs of another functional necessity. The discussion that follows identifies generic technologies that should support building services for a wide range of office building applications.

## Architecture

Within architectural disciplines many concepts must be applied to meet building functional performance objectives. In summary, these include raised flooring, open office planning, natural lighting and passive solar, low maintenance materials, and special function space to accommodate "nonwork" activities. There are also design and construction diagnostic procedures that must be applied to assure expected performance.

### RAISED FLOORING
The use of raised flooring, 8 to 18 in (20 to 46 cm) throughout office space is seen as a basic requirement in new construction projects. The wire management problems of the future simply demand it. No other system offers the degree of flexibility and accessibility to cable, and no other system of-

fers the capacity of handling the amount of cable that is expected to come into the office over the next decade.

All other floor system designs are flawed by their inability to allow access, proper capacity, and identification of routed cable. Our post-occupancy evaluations tell us that if you cannot follow wire to establish service, it will be assumed to be needed for some reason. Hence, when new connections are required, new distribution will be installed, abandoning existing cable. Over a period of a few years, cable will choke most conduit and cable paths of traditional systems. Another observation is that if tenants need to adjust service outlets, they must be able to do it themselves. The amount of time and expense to an electrician to pull wire, cut flooring, and install boxes is simply unacceptable for most tenants. As evidenced by many buildings today, cable connections are effectively fixed with occupants running taped and trip guard wiring everywhere.

The perceived additional cost of a raised-floor system is a false issue. If there are no other solutions to meet the functional need, there are no alternatives with which to compare costs. Even when compared with other cable management systems of reduced capabilities, such as floor duct, the costs are not that different. In fact, there are instances where raised flooring has been shown to be lower in initial cost than certain floor duct designs. Over a floor's life cycle, raised flooring dominates over all current technologies, even the low first-cost options such as poke-thru'. When building tenants move frequently (as much as once every 2 to 3 yr for the federal government), raised flooring becomes the lowest cost option by far.

One may also wonder about future power and signal transmission. If communication technologies evolve as presently conceived, using laser and radio frequency media, some type of enclosed pathway will be required: Raised flooring offers that path. Hence, the raised-floor concept appears relevant for both near- and long-term technology applications.

The major concerns with raised flooring are to apply durable materials and to ensure a good quality control process during installation. The use of cementious filled panels is an appropriate option for office areas as improved strength and acoustic properties can be achieved. Where heavy traffic exists, such as in corridors and pathways for hand trucks and robotic traffic, specifications must address requirements of high dead and rolling loads so that deformation is avoided. There is also a concern as to whether flooring panels should be secured. Using panel fasteners offers extra strength but inhibits easy access. In most cases, panels should be secured only in heavy traffic areas. Panels in office areas should be designed for easy lift out to assure use by tenants. During construction, flooring installation must be closely monitored, perhaps using leveling techniques such as laser sighting. If this is not done, even the best flooring materials may prove a disappointment.

Existing buildings pose difficult problems for flooring systems. Unless the building is being totally renovated, raised flooring is sometimes not practical because of accessibility problems of floor transitions and the impact costs of reworking architectural features such as windows, doorways, and elevator entrances. Most existing buildings have either ceiling-wall distribution or some form of floor duct. In such cases, perhaps one of the better solutions available is to partition office space into 250 to 300-sq ft (23 to 28 sq m) areas, enabling wire distribution to be permanently placed near possible workstations. Depending on the status of the occupant(s), the partitioned room could accommodate from one to four individuals. With four prewired utility service locations within such rooms, virtually all possible space use could be accommodated. This expensive approach is needed to offer the same space flexibility as raised flooring.

## OFFICE PLAN

Open office planning is being pursued for its space flexibility, energy savings, and cost impact to mechanical and electrical system design. Space flexibility comes by enabling an infinite variance in workstation location and configuration. Building system cost savings come because of current luminaire performance characteristics: with a parabolic type fluorescent fixture (two tubes every 64 sq ft [5.9 sq m]), a uniform, maintained 50 ft-c can be achieved at a power density of only 1.5 w/sq ft. Closed office space would have to increase power consumption to at least 2.0 watts to offer uniform lighting. Without special lighting control, such as occupancy sensors, the difference in annual energy performance between a building designed for open office and one designed for closed plan could be as high as 10%. There are also the first-cost differences of sizing HVAC systems and associated power distribution to meet the additional 0.5 w/sq ft.

This is not to say that closed office space is not desirable. Closed space does offer advantages with issues such as acoustics and personal privacy. It can even be speculated that operating cost differences might approach open space performance with the use of occupancy sensor lighting control, enabling a diversity factor of human presence to lower the effective power usage. However, the first-cost commitment to HVAC and lighting system design (based upon peak loading), coupled with loss in space flexibility, still swings the balance to open office planning. But, with continued advances in lighting technologies, fixture performance may eventually be less influenced by partitioning. In that case, the benefits of closed office may redirect our approach to office planning.

## PLANNING MODULE

Regardless of whether open plan or closed plan is used, a "planning module" is typically applied to a building. This imaginary building grid serves to align partitions, flooring, ceilings, and distribution terminals to minimize interference among interior features. A 4 × 4 ft (1.21 × 1.21 m) planning module has several advantages considering current lighting and ceiling technologies. As addressed above, this grid allows adequate lighting with a fixture spacing every 64 sq ft. The 4 ft grid also allows the use of the standard 2 × 4 ceiling panel, which is less in first cost than 5-ft (1.52 m) panels. An alternative 5 × 5 planning module usually results in fixtures every 100 sq ft (9.3 sq m), which eliminates the possibility of uniform lighting at a low power density. Only with future improvements in lighting system performance will a 5-ft module help meet both energy conservation and space flexibility objectives. This is a good example of how interrelated building decisions can be in providing functional needs.

## SPECIAL SUPPORT SPACE

Through use of special function space, there are opportunities to reduce maintenance and cleaning problems while improving aesthetic and functional qualities of the work environment. In all office buildings there are coffee pots, vending machines, and paper reproduction rooms. These messy areas have unique functional problems relating to odors, noise, and cleaning that can best be addressed if localized into convenient service areas for the occupant's use.

This principle can also be applied to the consolidation of file cabinets and the use of closets for personal effects. Centralized filing should be planned for every organizational element to get that function properly located outside of the workstation. Coats and other personal effects might also be better located in secured closets, eliminating worries of theft while improving space appearance by not having visible coat racks.

## FINISHES

To address the functional issue of low maintenance, there are a number of interior surface treatments that the architect can apply to reduce associated operating costs while improving appearance and acoustic properties at the same time. There are now available heavy fabric-type wall materials that improve appearance, acoustics, and wear resistance simultaneously. It can also be shown that, most surfaces made for flooring offer the potential of improving the durability of wall finishes. Special attention, however, must be paid to adhesives and use of fasteners as the hanging weight of these materials makes their mounting more critical. Flooring surfaces can be improved either through replacement convenience or improved resistance to soiling and wear. For most office space, the use of carpet tiles offer ease of replacement should severe soiling or staining take place. With raised flooring, carpet tiles are essential to assure proper access and acoustic properties.

## EXTERIOR ENVELOPE SYSTEMS

Exterior material selection also requires special attention to assure durability. Climate and physical abuse are the chief agents that will defeat exterior materials. However, the use of dissimilar materials is a recurring problem: For metals, oxidation and galvanic actions can take place. For other components, different thermal expansion properties can cause joint failures. All of these call for the thoughtful attention to material selections, configuration, and type of fasteners. A special review effort during design working drawings is required by the design staff to focus attention on the proper use of materials.

Moisture control is directly tied to exterior material durability and its thermal performance. Moisture migration and freeze/thaw actions are a major concern for northern climates. As fasteners and reinforcing become wet, spalling and staining of facade materials can result. If insulation becomes wet, either through direct leakage or condensation of vapor, its performance will be reduced dramatically. For these reasons, effective envelope vapor and air barriers must be used in colder climates, especially where humidity control is applied.

Wetting of roof insulation is a major problem for low-sloped roofing. In research by the Cold Regions Research Engineering Laboratory, extruded polystyrene insulation was found to be one of the few materials tested to withstand saturation over long periods of moisture exposure. Manufacturing claims for certain types of cellular glass insulation also speak to moisture resistance. Roof durability and thermal performance should improve with these insulations, particularly with protected membrane (IRMA) or single-ply membrane applications.

Further research has shown that the hot bituminous coatings of built-up roofing can cause outgasing of insulation materials, resulting in eventual separation and blistering of layers. Although not without installation problems, single-ply membranes do not use hot coatings, and as such, may offer a performance advantage. It has also been shown that roof gravel ballast can cause heat buildup due to light scatter. These two latter observations tend to suggest the use of non-ballast-type, light-colored, single-ply roofing membranes.

Experience has also shown that roofing reliability relates simply to getting water off as soon as possible. The use of steeply pitched roofing offers that feature. While low-pitched (flat) roofing is appropriate for roof-mounted equipment, there is no reason why the rest of the roofing surface cannot be sloped for rapid drainage.

Ultraviolet radiation from the sun can also create degradation of materials, particularly painted materials and joint compounds. The obvious treatment is to select those materials either naturally or especially made to resist exterior conditions. Materials requiring exterior painting should not be allowed. In many cases, coloring should be derived from natural pigments found in the raw material.

The combined effects of material compatibility, moisture control, and resistance to ultraviolet radiation lead one to the consideration of wall materials such as stainless steels, glass, and precast concrete panels. There are no hard and fast general rules for exterior treatments, however, as there are too many differing types of construction in too many climates and localities.

## ENVELOPE PERFORMANCE

An aspect of architectural design that has been previously undeveloped is the assurance that what was expected in design was actually delivered in construction. Joints are assumed to be tight, and insulation is presumed to be applied effectively. Our research with the National Institute of Standards and Technology shows that one of the reasons why buildings do not perform as expected is because they are not built as expected. The thermal integrity of building envelopes is extremely critical, for if we are not sure of delivered thermal performance we cannot be sure of environmental control. And indeed, buildings can be built so poorly that the designed mechanical systems cannot heat them. The need is to incorporate minimum levels of performance in construction specifications to be verified by diagnostic techniques such as infrared photography, pressurization, and gas tracer chromatography. With such techniques available to construction supervisors, thermal defects can be detected before the contractor leaves the site, assuring a better quality product and performance as predicted.

There is also a need to ensure that design assumptions relate to the real world. Thermal resistances have been measured in properly constructed buildings with indications that performance was 15 to 20% less than what handbook calculations would project. Another design assumption allows that building pressurization is effective in eliminating air infiltration: this is simply not so. Based upon such research findings, adjustments to design and construction criteria must refine our approach and expectations. Should the design standards community not update its reference calculation data and procedures, designers will be forced to apply "fudge" factors of safety.

Natural lighting and passive solar glazing treatments are closely allied and must be carefully designed. It has been said of solar energy that it is too weak to be useful as an energy source and too strong to not be controlled. Natural lighting has certain applications, particularly in mild climates, but only if used intelligently. Assumptions that furniture placement and glare are not limiting issues will defeat many natural lighting schemes. Past applications of sensor control have proved unreliable. Until natural lighting can be shown to be effective in delivering useful energy in a controlled manner, it should be restricted to noncritical applications. On the basis of our experience, warehouse, atria, cafeteria, and corridor uses seem appropriate, while office space applications require extreme caution.

This does not mean that passive solar treatments are not important. Building glazing and massing should be proportioned to take advantage of thermal benefits. Solar control through window insets, overhangs, and shading systems should be applied to take advantage of seasonal azimuth angles of incident radiation. Other passive solar treatments must be applied on the basis of the opportunities and limitations of site, climate, and occupancy. Regardless of how well the passive solar feature is designed, however, there should always be manual or automatic control over incident radiation. If this is not done, it is a certainty that space thermal control will be a problem.

## Interior Design/Furniture

Office planning of interior space and finishes has a dramatic impact on workstation and building system performance. It is here that all promises of occupancy loadings, acoustics, finishes, and space use are either fulfilled or broken. Before now, interior design services were seldom required as there was little to influence. The occupant had almost total control over interior environment. With Advanced-Tech space, the interior must be handled by a design professional versed in the interactive impacts of occupancy and system operations. There must also be an opportunity for the space planner to provide information to the design architect/engineer, assuring adequate system performance.

### BUILDING SYSTEM IMPACT

Without close coordination of colors, acoustic properties, fire ratings, soil resistance, etc., there can be unnecessary compromise in system performance. Surface reflectance percentages must meet lighting design assumptions. Without assurance of minimum conditions, lighting will simply not deliver required footcandles at workstation. And, as addressed in the architectural section, the issues of acoustic properties and durability of surface materials are significant functional interests directly related to interior design. Also, if improvements are to be made in fire safety and indoor air quality, interior design must address fire resistance ratings and outgassing characteristics of interior components. These concerns must be integrated into an overall compatible solution, involving all design disciplines.

### FURNITURE

Once past performance issues of finish selection, interior design relates to furniture; what to use and where to put it. Furniture has been a long-overlooked component of interior design by the government. It was generally concluded that office furnishings were the responsibility of the occupant. Unfortunately, many government furnishings are between 15 and 25 years old, space inefficient, functionally inappropriate, and aesthetically undesirable. The problem can be compounded as many interior designs are done with tenant personnel, often without professional expertise. Of increasing concern are the evolving requirements of office automation, reduced space allowances, and design integration of lighting, power, communication, and thermal comfort systems. Many of the related problems can be resolved if furniture is purchased and installed as a building system.

Furniture needs to be treated as an integrated building system to assure compatibility with other building systems and services. Its coloring and height will either support or defeat the distribution of lighting and air-handling systems. Its acoustic properties will either make or break an open office concept. With a program need for reduced space assignments, if furniture is oversized, there can be severe congestion of work surfaces. As the workstation is becoming the terminal of building systems (such as electric power, voice and data communications, and perhaps HVAC), there must be assurance that the completion of these systems involves the final point of connection. And, if the workstation is to support office automation operations, it must be ergonomically designed for such things as keyboard entry and computer screen viewing.

The use of "systems" or "modular" furniture is recommended for its compatibility with the issues addressed above. It is downsized (by about 20%) to offer improved flexibility and space reduction. Its height can be kept in the 4 to 5 ft (1.21 to 1.52 m) range to minimize lighting and air distribution problems. It comes with cable management cuts and channels to run power and communication wiring. It is also ergonomic in design to allow easy keyboard entry to computer terminals. The furniture's only problems relate to time-consuming assembly and acoustic privacy.

Once a building is fitted with such furnishings, it will be necessary to ensure that it remains functionally complete by ensuring that future tenants have the same type of units as provided to the original tenants. To achieve this objective, professional assistance is required in the management of the system to assure proper use and servicing. For this purpose, a multiyear services contract should be considered part of the purchase price.

## Structural

There are several innovative structural design features that can be incorporated into an Advanced-Tech building. Some are limited to specific applications relating to seismic conditions, corrosive environments, or extreme building height. In summary, innovation can be seen in the use of epoxy-coated reinforcing, high-strength concrete, base isolation support systems, and diagnostic monitoring.

### EPOXY COATINGS

Epoxy-coated reinforcing steel is particularly useful in avoiding spalling of cast concrete members. The coating does not allow moisture contact with the reinforcing steel, thus avoiding oxidation and its associated expansion strains. The coating does not compromise structural integrity. This type of treatment is particularly important for construction near coastal areas, where the atmosphere contains salt vapors that accelerate the corrosion process of reinforcing steel. This technology is an extension to the current use of epoxy-coated reinforcing in bridge construction and roadways subject to salt deicing.

The availability of new systems does not necessarily mean that adoption is appropriate. The application of post-tension structures is an example. This system takes the place of concrete reinforcing bars by tightening continuous rods that pass through tubing within floor slabs: The resulting compressive loading offers required strength to the concrete deck. Unfortunately, these rods cannot be cut without structurally weakening the building. This restriction could make floor penetration retrofit difficult and limiting.

## HIGH-STRENGTH CONCRETE
Concrete with ratings of 6,000 psi and above is attractive in high-rise applications where the structure can create large column loadings. By the use of such concrete, building weight can be reduced, thereby achieving cost reduction and improved space efficiency.

## BASE ISOLATION
The concept of base isolation structural systems is particularly attractive in high seismic risk areas for low- to mid-rise structures. By providing base isolators between each column and its footing, the isolators allow the earthquake movements to be independent from the building, filtering the seismic forces before going into the superstructure. This is very important as building codes do not provide for the concerns of building interior survival; they being centered instead upon the objective of avoiding building collapse. Building contents may be totally destroyed after a quake unless special attention is given to their protection. Base isolation offers that protection with virtually no increase in construction costs. Special coordination efforts need to be made to assure that building utility connections can accommodate building movement. There are also concerns to assure that architectural site development in no way restricts building movement.

## DIAGNOSTICS
Finally, there is need to better understand how our buildings react structurally to external forces, such as foundation settlement and seismic motions. By instrumenting key structural members with strain gages, it would be possible to monitor structural integrity over time. The information from the instruments could be fed into the centralized "diagnostics center" discussed later under the mechanical section.

# Mechanical

The concerns of building energy performance and precise automated control of environmental conditions are major functional issues in treating mechanical systems. However, there are also the concerns for space flexibility and air quality. Most of these are addressed by providing state-of-the-art system designs, intelligently configured and zoned for control flexibility.

### ENERGY EFFICIENCY

The use of energy-efficient equipment is a given for virtually all mechanical equipment. It is interesting to note that equipment today was simply not available during the energy "crisis" of the 1970s. We can now consider boilers with efficiencies between 80 and 90%. Chillers can be easily purchased with peak load efficiencies offering 0.6 to 0.7 kw/ton. Duct designs are becoming virtually airtight. Fan systems are being configured for lower static pressures. And, all are being better controlled with standard optimization strategies. Some manufacturers are even standardizing heat-recovery and free cooling options in their product lines. The only challenge to the designer is to ask for them.

### DIRECT DIGITAL CONTROL

This type of control system has been actively marketed for only the past few years. However, with current trends, it will eventually replace pneumatic and electric controls. Direct digital control (DDC) is now less expensive, and offers the promise of improved system performance through more precise control. The concept centers around the idea of having equipment controlled by sensors that "talk" directly to a computer-based automation system. This digital signal is claimed to be more accurate and avoids transducement losses. It also facilitates the use of powerful building automated systems (BAS) as it allows low cost signaling of input, remote monitoring, and reset functions. Such systems, however, are now without standardization and once committed to a manufacturer you are effectively locked in. To support the needs for reliability and to avoid price gouging down the road, the use of multiyear service contracting should be made part of the construction contract: such as an initial 1-year requirement with four 1-year options to renew at a predetermined rate.

### BUILDING AUTOMATION SYSTEMS

It is difficult to relate to this concept as technology is changing every few years. Most experience is with older systems, and many of us have not been reeducated since the first systems were installed in the mid 1970s. Twelve years later, we have seen about four generations of BAS technology, the latest being the integration of control signaling through a building-wide telecommunication system. The old argument that BAS systems are too complex to run is not without merit but there seem to be few options if one is interested in higher levels of building efficiency and energy performance. As with the DDC discussion, the use of long-term contracts are absolutely essential to assure continued performance and accountability. However, it is clear that buildings management will have to upgrade its operations to address this technology, perhaps to the extent of recognizing that an electronics engineering technician classification should be established to support advanced computer-based control systems.

## BUILDING DIAGNOSTICS

Related in function to a BAS is the development of a diagnostics center in each prototype building. This center will monitor critical environmental and performance features of the building. Air quality will be one of the most critical monitoring functions. While we do not believe that PBS buildings suffer air quality problems in general, there may be transient instances where air quality should be improved through increased ventilation: such as when new finishes are applied, during office moves, major space alterations, or cleaning operations. The diagnostic station should also be able to monitor ventilation exchange rates and to profile the building's outdoor air use. Providing sampling tubes throughout the building for tracer gas measurement (such as sulfur hexafluoride) will greatly facilitate these measurements and could be used over extended periods to get better year-round data on building air use and leakage.

In effect, our buildings should become laboratories and developed into case studies to teach us about our inventory and how it should be run. If we are to react with certainty, we must become quantitative in our assessment of cause and effect relationships.

## SYSTEM CONFIGURATION

Building systems and heat-recovery schemes will suffer if they are not configured and controlled to meet concerns of redundancy, zone control, and localized off-hour service needs. This usually means that air handling must be compartmented to allow at least floor by floor conditioning, avoiding vertical zoning. Configuration designs also require proper load splitting between multiunit plant equipment. In most cases there will be a need for a stand-alone heating and/or refrigeration source for off-hour loads, separate from the needs of specialized load space such as computer operations.

Configuration also calls attention to the need for redundant components to assure that critical operations can be maintained in the event of system component failure. While previous comfort air-conditioning design would not provide standby chiller or pump capacity, we must now be sensitive to the needs of office automation equipment and to the concerns of worker productivity.

## BUILDING VENTILATION

A recurring problem with buildings is their use and distribution of ventilation (outside) air. We use too much or not enough or don't put it where needed. This is particularly true with variable air volume distribution systems as outside air is difficult to proportion to varying air supplies. There has also been the observation that ceiling supply discharges can short cycle over the workstation to commonly placed ceiling returns. To meet this problem, low side wall returns would draw supply air downward. Another solution would be to provide a dedicated ventilation air system designed to deliver minimum ventilation needs directly into the workstation. There have even been some German designs that provide for an air terminal to come up within or alongside the workstation desk, facilitating occupant control.

## TASK AIR-CONDITIONING

To offer possible improvements in energy conservation and comfort control, there is a concept of task air-conditioning. As presently evolving, such a concept provides constant volume, conditioned air through raised-floor panel fan units, using ceiling return to a central station air handler supplying low pressure air to the floor plenum. Relatively large quantities of air (about 1 to 1.5 cfm/sq ft) are required to assure that floor discharges are not too cold for leg comfort. However, if interfaced with occupancy sensor controls, as discussed in the electrical section, such a system might rival variable air volume economy because of the diversity allowance of not having people at their workstation. There are also the benefits of assuring improved circulation in the workstation by injecting directly into the occupied zones of the room. With current code requirements, however, such use of a raised-floor plenum requires that wiring within the raised floor must either be in conduit or be insulated with fire-resistant Teflon. This could prove limiting because of high cost.

## HEAT RECOVERY

Thermal storage, free cooling, and heat recovery are all well-established principles of energy (or dollar) conservation. These should always be applied if supported by an engineering/economic analysis. Although thermal storage is not always cost-effective, there are many proven applications of free cooling and heat recovery. Free cooling through an air economizer cycle is practically a control standard. Free cooling using condenser heat exchangers is also considered commonplace today. Heat recovery is becoming more widely used, but usually requires costly equipment, making it less attractive than free cooling. One heat-recovery application that has been shown to offer significant savings is the transport of mainframe computer room heat to meet building perimeter heat loss. There is also the easy heat recovery of interior zone loads by using a ceiling plenum perimeter separation with ceiling-mounted fan coil or fan-powered VAV units, capable of moving warm plenum air to the perimeter. Another simple application of transfering interior heat gains to the perimeter is through zoned heat pump air-handlers, interconnected through condenser distribution.

## PLUMBING SYSTEMS

Plumbing systems offer innovation primarily in reducing water and energy consumption. There have also been advances in better understanding sizing requirements to avoid unnecessary first costs. For some time, PBS has favored the use of single temperature faucets, delivering 100° to 105° F (38.7° to 40.5° C) water to lavatories. By not providing the extra distribution of a two-pipe hot and cold system, there are obvious first-cost savings. And by not operating a hot water system at 140° to 180° F (60.0° to 82.2° C), there are heat transmission savings. The concept has been proved and will be continued as a design standard. Should the building require food service—related dish washing or other high-temperature water service, there will be a need to consider the use of multiple hot water heating systems.

The use of flow-restricting spray nozzles for faucets and low-flow flush-meters for fixtures is commonly accepted. However, there is now being marketed a line of fixtures for even lower flow rates, using no more than 1 gal (3.78 l) per flush. To further conserve water, one can investigate the application of "gray water" systems, to use water more than once before throwing it away. Instead of sending lavatory sink water to waste, it may be cost-effective to consider treating it for a second use as toilet supply. There is also the more common application of using rain water and gray water effluent for site watering. As site watering accounts for more than half of nationwide use, this application may dramatically reduce building water consumption. There needs to be better industry support of this concept in terms of manufactured products and code development.

In terms of code development, the National Institute of Standards and Technology has evaluated the performance needs of drain systems and found them to be overly conservative. Based upon this work, new pipe sizing criteria are being considered for our prototype buildings as a means of reducing first and maintenance costs.

# Electrical

There are significant innovations in the application of lighting and power systems. Most meet the functional objectives of efficient and reliable service while supporting the needs of office automation. There are also control and monitoring strategies that should help reduce energy and utility costs.

## OCCUPANCY SENSORS

Lighting control represents a major opportunity for utility savings. With today's technology, the best way of improving lighting efficiency is to simply define the conditions under which lights should be on and off and regulate accordingly. In most office applications, lights should be on when there is someone in the room and off when there is not. Since experience shows that we cannot rely on occupants to manually control their lighting, this optimization requirement can only be met through automatic control, using an occupancy sensor. No other means will offer the flexibility, convenience, and assured cost savings. Control by BAS will not offer the same degree of savings as it will typically regulate lights between two control time periods, early morning (on) and late night (off): This time clock control will miss all the opportunities of saving during the day when a room is vacant. As this can be as much as 30 to 50% of the working day, there is considerable advantage in choosing occupancy sensor control.

Lighting sensors work best in closed office space but can also be applied to open planning. In an open plan configuration, a much broader area of control must be used since the amount of light delivered to the workstation depends on the adjacent space fixtures and associated background scatter. Even in the open plan, however, the control performance of occupancy sensors is much better than that of other means.

Both infrared and ultrasonic occupancy sensors are now recognized for their cost-effectiveness and acceptable reliability. Cost-effectiveness has been proved in most applications involving moderate to expensive electricity. Cost-effectiveness is improved where wall switch costs can be avoided, relying solely on the sensor for control.

## CONTROL INTEGRATION

Occupancy sensors can also expand our scope of automated services to include status indications for security and fire safety use. If occupancy sensors can be prompted to indicate where people are in the building, the application is clear as to their value in identifying intruders or occupancy of secured space. Fire safety evacuation procedures could also be enhanced by their noting precisely where people were in the building who might need assistance.

## LIGHTING SYSTEM

To optimize energy and cost savings further, the lighting system must be efficient itself. The most efficient lighting source today, which is suited to office space color needs, is the parabolic luminaire with the T-8 fluorescent tubes. As addressed in the architectural section, when two-tube parabolic fixtures are spaced in an 8 × 8 ft grid, an open office plan will allow 50 ft-c to be achieved almost uniformly throughout the space, while using very low amounts of energy. This even allows for a 15% derating of tube performance, on top of the maintenance factor reduction and a reasonable obstruction allowance.

There are other functional advantages to parabolic fixtures. They provide very little glare, so they are well suited for computer terminal environments. And, with the offering of uniform distribution, space planning is made more flexible since it makes no difference where desks and other work surfaces are placed.

However, experience has shown that care must be given in the specification of parabolic fixtures, as there are some manufacturers whose fixtures are poor performers. Specifications must address desired illuminance levels, power density limitations, tube lumen output, ceiling heights, room cavity ratios, surface reflectances, etc., in order to assure that the delivered fixture will be as efficient as expected. But, with all the variables at play there is no proof of performance without an actual test. On major projects, it is recommended that a mock-up of new spaces be provided to validate performance expectations.

## POWER DISTRIBUTION

Electrical services to wall or floor outlets require new directions for the automated office. At issue are concerns of relocation flexibility and suitability of service from the standpoint of assuring "clean" and reliable power to office automation equipment.

For new construction and large-scale space renovations, a raised-floor system should be favored to allow ease of cable installation and space flexibility. Underfloor junction box plug-in flexible connections to receptacles should also be used to allow quick disconnect and reservicing of a moved workstation. For minor renovations to existing office space, the addressed material in the architectural space planning section would call for going beyond normal wall distribution/placement, offering enough outlets to service any perceived occupancy/configuration.

The automated office should also have two separate power distribution systems, one for noncritical functions such as pencil sharpeners, desk lamps, etc., the other being designated solely for office computer services. This separation is economically justified to minimize power outages associated with occasional space alterations and overloaded circuits. With a separate distribution system, there can also be entertained an uninterrupted power supply (UPS) to certain critical workstation computers. A separate power system, if sourced from a dedicated secondary transformer, also helps limit power line voltage noise from building equipment such as air-conditioners, elevators, and paper-copying equipment. This is critical as power line noise can cause computer equipment failure. Even with these features, power conditioning units should be considered for individual computer connections, avoiding noise transmission from computer to computer. These concerns are functionally important to ensure reliability of service to the workstation. If the automated office demands the use of electronic equipment, the building must meet those associated needs of flexibility, compatibility, and reliability or fail in its most basic service to its occupants.

## POWER MANAGEMENT
In support of BAS control and monitoring functions, the Advanced-Tech building should be extensively metered to allow better awareness of what is using how much power. Each major piece of equipment and power panel should be fitted with an electric meter. There may also be the need to separately track utility consumption for billing purposes of joint-use/commercial space, mainframe computer operations, extended-hour operations, or other specialized areas.

In conjunction with utility metering, demand control will probably be appropriate wherever there is time of day billing. The problem is to identify those loads that can be terminated to satisfy limit control. In some applications, office services will be considered so critical that demand control is not possible. In these cases, the use of nighttime thermal storage for air-conditioning refrigeration may offer one of the few means of cost-effective downloading of daytime power consumption.

## TELECOMMUNICATIONS
With the divestiture of American Telephone and Telegraph (AT&T) and the deregulation of the telephone industry, it is now possible to become more involved with intrabuilding telecommunications. As tenant agencies and the GSA Information Resources Management Service will typically control procurement of telecommunication systems, it is critical to effectively coordinate building equipment network needs for automation functions. With such an approach it is also possible to entertain the idea of new technologies' involving integration of building communication services for both office and building automation.

## SYSTEM INTEGRATION
It is now technically feasible to integrate virtually all building voice and data signal requirements into a common telecommunications system. Such offers the ability to improve relocation flexibility and to lower first costs by avoiding unnecessary distribution of signal cabling. One possible configuration uses a digital switch (PABX) to effectively integrate communications into a star network with the PABX as the central "brain" of the building. Other configurations would make use of an integrated cable network, providing designated wires or frequency band widths for each type

of communication service. In either case, the flexibility feature, coupled with office automation capabilities, has supported terms of *smart building, intelligent building,* and *office of the future.*

## SYSTEM COMPONENTS

With signal equipment relocation or signal redirection, the PABX can be reprogrammed to address the new location. Such switching is appropriate if line use is not extended for long periods of time. The design of the PABX must be sensitive to expected traffic of voice and data, less there be the potential "blockage" of signals. In the case of volume data transfer, such as might exist between larger computers, current technology would suggest the use of a local area network (LAN), offering dedicated lines for that purpose.

Cable distribution options are numerous, and selection will be heavily influenced by requirements of signal transmission bit rates and the design requirements of switch and instrument interfaces. Although fiber optics is now considered cost-effective in many applications, such as trunk distribution, the application of twisted pair still seems to dominate current system development. It seems likely, however, that fiber's improved capacity, resistance to interference, and eventual low cost, will establish a market as the preferred medium of signal transfer.

Facility support of telecommunications will involve special attention to space assignment, environmental control, and physical security. Basically, telecommunication space is becoming aligned with computer rooms when treating environmental and security needs. Mechanical and electrical supporting systems generally require dedicated sources to assure year-round service. Such systems also require redundancy to improve reliability of communication services. Floor space dedication is generally increasing because of integrated voice and data concepts. Not only are the sizes of switch rooms and communication closets increasing, but in terms of distribution requirements, space provisions must be made to assure an adequate amount of vertical passages between floors. Use and placement of sleeves is becoming critical in many buildings. Space planners and designers must become more involved in assuring proper space needs of future telecommunication expansions and changeover requirements.

With the provision of the switching center (PABX) and the distribution of cabling, there is a requirement to assure that tenant agencies are supplied with the necessary instruments to make the system work. If instruments are not coordinated with the overall telecommunications procurement, there is risk that scheduling problems will result in having people move in without communication services. There is also a possible compatibility problem as some instruments may not work directly with the rest of the telecommunication system or be limited in accessing the full range of functional services. Voice (and perhaps data) instruments must be made part of a telecommunications system design. The type of instruments provided should meet both current and projected future needs to ensure functional flexibility and support.

## FUNCTIONAL CAPABILITIES
The capabilities of the building's telecommunication system will largely depend on technology limits at the time of procurement. Consistent with concerns of applying systems suited to the basic mission of the tenant (and meeting issues of data security and system reliability), communication systems should be the most technically advanced with as many functional options as possible. When functional services are provided, instead of, "When in doubt, leave it out," the applied principle should be, "When undecided, make sure it's provided." The first costs are so insignificant compared with productivity issues that it is better to "force" a tenant to deal with a system offering greater capabilities than to target a system to meet current needs that might limit a tenant's mission accomplishments in the future.

## BUILDING AUTOMATION
As addressed in the "Mechanical" section above, BAS interface through the telecommunication system is now a possibility. This application takes several forms depending, in large measure, on the manufacturer of the communication/control system. To some, integration of building automation involves the shared use of telephone cable for both office automation and building automation signaling. To others, building automation can be supported by the PABX to switch incoming and outgoing signals. There is not much value to interface through the PABX unless signals are to be routed to different locations. In the case of most BAS equipment, there is little expectation of relocation and as such little value to interface through the switch. Most BAS applications can best be served through a local area network dedicated to building functions.

## SERVICING
As with other sophisticated system designs addressed thus far, service contracting appears to be a necessity in order to assure technical support, accountability, and fast service. Five-year contracting is recommended as a minimum. The construction contract should start the process by calling for a first-year service agreement, with a minimum of four 1-year extension options. The requirement for multiyear services is necessary to ensure continuity of support and to allow for long-term shakedown of all operating components. Such contracting is necessary even if a building operating staff is available: the multiyear servicing of equipment will ensure adequate familiarity of system operation long after postconstruction training would otherwise be forgotten.

## TELECOMMUNICATION CONTRACTING

Whenever building system integration takes place, there are the potential problems of nonperformance accountability and misscheduling of deliverables. As it is typical that HVAC, fire safety, security, and building diagnostics systems be procured using fixed-price procurement, their associated BAS control should follow suit. Unfortunately, most telecommunication systems are now being purchased on the basis of a request for proposals. There is no compatibility in these two procurement methods. Policies and criteria must be pursued to help establish procurement guidelines. If telecommunication service is procured as a building system, the design and construction coordination needed to assure system integration of equipment can be assured. If the procurement of intrabuilding telecommunications is divided by contracting sources and/or methods, extensive efforts will be required to administratively assure equipment and scheduling compatibility.

## FACILITY SUPPORT TO OUTSIDE NETWORKS

A major feature of communications will be its outside connection to intracity and intercity networks. In addition to underground cable connections to tariffed lines, in the future, virtually all buildings will sprout antennae and dishes of almost every description. It is important to recognize that there are important facility impacts to design disciplines when providing such services. Wind loads could represent structural concerns for roofs with large antennae. Architectural placement of antennae and the building itself must also be sensitive to azimuth angles of satellite communications. This can sometimes have an impact on roof design for cooling towers, elevator penthouses, and other roof structures. Such should also influence site selection, building placement, and configuration.

## TELECONFERENCING

A new and evolving telecommunications function is that of video teleconferencing. Instead on relying on audio (telephone) communications and/or paper transmittals, organizations can apply two-way visual contact to achieve a more personal and effective exchange of information and direction. Of the several forms of teleconferencing available, the most functional is live action two-way video conferencing. As current technology limits networking through twisted pair, the high-speed transmission requirements are effectively limited to coax, fiber, microwave, and satellite media. For networks with interstate capability, the most common configuration currently is satellite transmission.

The limiting influence in adopting this new function primarily relates to high operating costs. In order to cost justify the expense of video teleconferencing, agencies must experience a high volume of information exchange that otherwise would require extensive travel and/or contact with organizational units. In the case where multiple tenants in a single

building have a collective interest that might justify video teleconferencing, there is justification for considering a concept of shared tenant services (STS). In order to make STS work, a major tenant or service
agency must accept operating and administrative burdens of establishing
network agreements and managing reimbursable accounts.

Video teleconferencing is typically evaluated for each new office building. Where tenant agencies can support such services, teleconferencing is
being applied as part of the building construction contract. Where such is
not initially justified, building design should configure at least one of its
major conference rooms to allow for eventual retrofitting.

### VOICE/DATA SECURITY

Unauthorized access to building information systems must be eliminated
as an issue if office automation is to be provided on a broad scale, and certainly if LANs are to be applied. There is also the worry that "system bust-
ers" (those outsiders who attempt to access computer files) may interfere
with office operations and security measures. To address these concerns,
access codes and data scrambling/decoding are security concepts that
must be approached by the building's telecommunication system. Perhaps
in more secure installations, there should also be programming such that
beyond a limit of unsuccessful entry attempts, an alarm condition would
enable a line trace for detection of source.

## Fire Safety

Although the government's conventional buildings have historically been
designed for high levels of performance in their resistance to fire, there
are new technology applications that can offer enhanced monitoring and
suppression capabilities. Fire safety comes with attention to minimizing
fire loads and sources of combustion while offering faster response and
more comprehensive applications of fire alarm and suppression systems.

### OCCUPANCY SENSING

One of the most innovative approaches to fire safety, the use of occupancy
sensors, has already been addressed under the concepts of lighting control
and building automation. The status of such sensors can be monitored
through BAS central console, shortening the time to mere seconds in determining if the building has been evacuated: such can be critical in large
buildings in which might take as much as a half hour to establish if an evacuation has been completed.

### SYSTEM INTEGRATION

As discussed earlier, alarm and sensing control features can be accommodated through an integrated building automation/telecommunication
system. At issue is whether the associated flexibility and reduced cost can

be realized, while providing traditional levels of reliability. The problem can be resolved through a distributed BAS architecture. It is now possible to apply fire safety as integrated subsystem with a BAS but with stand-alone capability should the BAS processing computer fail.

## FIRE SUPPRESSION
The use of wet pipe sprinkler systems throughout a building, regardless of story height, is seen as a way of improving life safety and sometimes of allowing a more nontraditional development of architectural features. The use of atria, open stairwells, and other unique features can be facilitated with the assurance of sprinkler protection. Although not essential for most office operations, the use of automatic self-closing heads can be considered to limit the amount of water involved in fire suppression. Fast-response sprinkler systems can also be applied to increase speed of activation, thus minimizing possible fire damage. There is also a possible option to provide floor and ceiling cavity sprinkler protection to help limit the worry and acceptable performance standards for smoke control and fire stop ratings between floors.

Special fire suppression systems, such as Halon, can be applied to critical areas where sprinkler water damage is of extreme concern. Halon systems also enable rapid detection and extinguishment, within seconds of combustion. The BAS and PABX computer centers offer new building applications. Limitations are in treating materials that create their own source of oxygen or pose deep-seated fire potential. Halon's high cost, toxicity and impact to the ozone layer precludes use in large volume or populated areas.

## FIRE ALARM
Fire detection systems are also being considered for the entire building to assure a quick response. The application of smoke detectors is particularly appropriate for locations that might support smoldering fires. It has yet to be determined whether smoke detection will offer a significant lead time between early detection and sprinkler activation, especially if fast-response sprinkler heads are used.

## SMOKE CONTROL
Associated with smoke detection, the use of smoke control systems is a very standard feature in high-rise buildings. Within certain limits, such are also appropriate in smaller buildings to assure clear stairwells and use of elevators. Experience dictates the use of separate smoke control fan systems, avoiding the control problems of dual function for building air handlers.

## FIRE LOADS
Among the criteria for furnishings addressed under "Interior Design" above, should be the use of fire-retardant materials. While these will not

eliminate fire loads, they will substantially minimize what has traditionally been a major source of smoke generation. Particular attention should be paid to free-standing partitions and cloth-covered office furniture. Carpeting must also be fire resistive to guard against dropped sources of combustion.

## Security

Advanced technologies relating to physical security typically employ established security measures to an expanded number of applications. As with fire safety, there are also the opportunities of exploring new monitoring functions and system integration to enhance system performance, ensure flexibility, and reduce costs.

### OCCUPANCY SENSOR
As with fire safety, the use of occupancy sensor lighting control could be a significant new feature for improved security monitoring. At issue is whether proper safeguards can be developed to allow the use of exposed telephone wire. If this proves acceptable, the system may be developed as a subordinated service, relying on more sophisticated and protected intrusion systems to meet primary monitoring functions.

### CARD KEY LOCKS
One application of a combined security system is the use of card key locks in place of traditional metal key locks. This approach offers improved service as an accounting can be assured as to who went where at what time. Lock changes can also be made through a central computer, avoiding the use of locksmiths. Such can be applied to virtually all room and closet locks. This is being done in the more modern hotels being constructed today. Not only will access be denied to unauthorized users, but a record of the attempt, with associated alarms, is available to identify where and when the intrusion was attempted and by whom. Integration of card keys with building picture passes could also reduce tedious log-in/out procedures. Card key control could be cumbersome if not handled properly in a noncritical office environment. Consequently, their use should be available to all doors normally fitted with locks, but featured with an electronic means of holding the lock open for periods of free access.

### CLOSED-CIRCUIT TELEVISION
Perimeter intrusion sensors and closed-circuit television is nothing new: both offer an opportunity to supervise critical means of entry and access. For occupant security, closed-circuit television can also be used to monitor parking lots and similar areas to assure safe passage. The problem with television cameras is that they offer occupants a negative statement of mistrust and give rise to a "big brother" syndrome. Cameras should be placed

conspicuously only as a visual deterrent: special concealment should be exercised for most internal applications.

## INFORMATION SECURITY

New security threats come with computer and electronic eavesdropping technologies. However, there are physical security steps that can minimize such intrusions. Physical barriers, especially those constructed of heavy concrete or shielding material, will limit outside contact through the perimeter. Music or white-noise systems directed toward an area's perimeter skin can also mask inside conversations from outside locations. Nonconductive inserts in metallic penetrations of space can also minimize signal transfer. The application of such measures must be weighted against the sensitivity of the information being processed.

# Robotics

An evolving set of technologies that are just beginning to impact building design is the use of robotics. The applications are similar in theme to most automated functions; that of doing mundane or precision tasks. Though applications are now limited, their expectations are unlimited. Our buildings must be designed to accommodate these machines with a minimum of conversion problems.

## MATERIAL HANDLING

In current use today are robotic mail trucks. Their operation is based upon a tracking system that is sensitive to special tapes or spray coatings. Such equipment could also serve other delivery functions such as paper reproduction pickup or other material handling. Presently, all require human attention for loading and unloading.

## WINDOW WASHING

A robotic service that is now under development is window washing. As presently conceived, the device would track up and down exterior wall guides, requiring little attention by human operators. It may be several years before this system is marketed, however, as the frequency of window washing limits the potential cost savings as applied to a single building.

## INTERIOR SERVICES

Floor cleaning is also an application that could have significant operational impact, particularly for large open surfaces. There are numerous other potential services such as security patrol, trash pickup, landscape maintenance, and food service preparation. All of these will require special attention to accessibility. Entrance thresholds will present problems, as will careless placement of plants, water coolers, fire extinguishers, and other obstacles. Turning radii must also be considered in providing architectural obstructions. For mobile robots, there must also be attention to assure that floor finishes are durable enough to withstand repeated localized rolling loads: this is particularly important for raised-flooring panels and carpeting. Design criteria, similar in detail to the government's accessibility standards, will likely be required.

## DESIGN AND CONSTRUCTION PRACTICES

Advanced-Technology concepts must not only relate to building features but to the design and construction process that delivers them. For without innovation in design and procurement tools, the traditional analytical, procedural, and operational processes of the past could limit our efforts in optimization. When seeking higher levels of performance, there must be the means of evaluating precise operational differences among alternates, of performing economic studies of more optimization alternates, and of assuring quality control to get a building that will operate as intended.

### Computer-Aided Design (CAD) Drafting

Past design practices called for architects and engineers to develop drawings and specifications in camera-ready form to support reproduction of construction contract documents. Contractors in turn used these reproductions in estimating takeoffs, in developing shop drawings, and eventually in updating field changes. Still other material such as shop drawings and catalogue cuts and operating brochures were developed and made part of what has turned out to be an overwhelming amount of useless paper. Building managers and space planners made little use of this material since it was usually in a form not adapted to their needs.

With the use of CAD drafting, this can change. With the selection of CAD drafting systems that are also compatible with facilities management (FACMAN) software, design and construction data can be stored in a useful format ready for operational services. Space layouts, inventory management, spare parts listings, preventive maintenance scheduling, location guides, and many other services can be directly generated through CAD.

The application requires that the design must be generated using CAD drafting, followed by its continued use in the field for as-built changes. Shop drawing and operating manual instructions changes can also be recorded along with a host of inventory and specification data. All of this transferred and accumulated data can then be readily retrieved and used through FACMAN interface by the building operating staff.

A limitation of its use is that most buildings management operations are now committed to, or are developing, their own data management systems. If at all possible, these two systems should be made compatible to allow ease of data transfer and extension of existing hardware. This is a problem that should be resolved over time.

Another application issue is that the system must be designed to accommodate other buildings that might be under the control of the local buildings management staff. It will do little good to have one building fitted with ultrasophisticated facility management functions, while 10 others charged to the same operating field are being treated some other way. To make it work, the system must have the capability of assuming all management functions of the manager's field office.

## Computer-Aided Design Analysis

CAD also refers to a growing list of computer software that can support design calculations associated with system selection and sizing. There are those that should be considered mandatory while others are preferred.

Overall building energy-use projections must use a computer-based approach to accurately account for climate, interactive loading, and part load system effects. HVAC and fuel selections are heavily dependent on CAD in estimating energy performance and accounting for time of day cost impacts. Without CAD, passive solar analysis could simply not be done credibly nor economically. There is also the mandated use of CAD in the analysis of such systems as base isolation, simulating the effects of seismic disturbances on the building's superstructure.

Although not mandatory, there are numerous software programs available to improve design accuracy and productivity. HVAC equipment selection, ductwork layout, and pipe sizing can all be done now by computer. Hydraulic sprinkler design, life cycle costing, lighting distribution point calculations, and structural member analysis are further examples. There are also management support programs for project administration and cost estimating. Although not required, the proficiency of their use should be an evaluation factor in the selection of architect/engineer firms.

## Construction Standards and Performance Measurement

Without a means of verifying design performance levels, there is no way of assuring that critical design operations will work. The issue relates to increased quality control measures necessary to commission more sophisticated equipment associated with Advanced-Tech features.

Diagnostic applications go beyond traditional construction supervision practices by including new techniques and testing applications. To be applied to future new construction projects will be requirements for envelope testing using infrared photography to identify thermal anomalies.

Also the use of gas chromatography will be applied to establish infiltration rates as applied against minimum performance specifications.

Through agreements with the National Institute of Standards and Technology, PBS will be working toward development of a comprehensive set of commissioning standards and field measurement techniques to assure mechanical equipment capacity and efficiency. Incorporation of these reference standards in guide specifications is in progress to support proper applications. Such inspection/commissioning practices are needed for all building systems.

## COST IMPLICATIONS

Although Advanced-Tech will raise construction costs over conventional building stock, the increase is not large compared with costs of "traditional" designs. Added costs may approach 5 to 8% for enhancements to building systems and features. Where basic voice/data telecommunication costs have been included, construction costs can increase an additional 2 to 5%.

Depending on material quantities/quality and occupancy distribution, an additional 10 to 20% increase can be expected if furniture systems are included. However, in most cases, this cost should be viewed as unavoidable and sunk. For if furnishings are past their useful life, the commitment to replacements is needed even if not associated with the Advanced-Tech project.

There are other cost increases in the delivery of Advanced-Tech space if design programming and architect-engineer service fees are included. As more is being asked of these design professionals, their reimbursement will increase. The amount should be roughly linear with construction costs. Only in the design of extremely complex building features, such as base isolation structural systems and integrated telecommunication systems, can it be expected that fees will increase above percentage changes in construction costs. For major projects, even with complex building features, design fees should be within the government's legislated limitations of 6% of construction cost.

## PRODUCTIVITY RELATIONSHIPS AND IMPLICATIONS

A basic objective of the Advanced-Tech program provides for improved tenant productivity through enhanced building features and services. A building's influence on productivity can be addressed in two categories, that associated with facility environment and that relating to provision of building equipment and services that directly interface with the occupant's mission.

It is generally recognized that facility-enhanced space offers improvements in occupant productivity. Such comes through assurance of thermal comfort, proper illuminance, acoustic privacy, workstation flexibility, ergonomically designed furnishings, minimized travel/distances, reliable power/operating services, and aesthetically desirable space. Unfortunately, little conclusive psychological research has been done in these areas to quantify cost implications.

Productivity impact can also come with provisions of mission-related equipment that is directly used as a tool by the tenant. The most recognized application is the use of computer services. Improvements can also come through robotic support and special telecommunication functions such as video teleconferencing.

When cause and effect relationships of advanced technologies on efficiency are addressed, the easiest to quantify relates to complete downtime of workstation activity. Even then, these analyses must be qualified by the use of probability estimates of likely occurrence and severity impact. For instance: What is the probability of an agency moving its staff, and how long will the disruption last? In this case, experience might support numbers such as once every 2 to 3 years with a disruption of 3 days for packing, moving, and unpacking). By relating to the lost hours of this nonproductive time, there is a functional relationship between moving and the lost time expense of not accomplishing the occupant's mission.

This leads to a problem in quantifying productivity impacts, or determining what the occupant's time is worth. The problem is compounded further since occupant worth might vary with time of year and even time of day. For the government's nonprofit service-oriented missions, it may be appropriate to generally assume that the value of the employee's time is equal to salary. Such could be done in the name of conservative estimating, but more honestly, is done because of our lack of ability to quantify monetary impacts of mission. But even without an exact accounting of mission-related costs, the salary cost impact of disrupted services is overwhelming compared with other building costs.

Consider the downtime of a workstation with an employee average take-home salary of $30,000/year. If we allow for overhead expenses (vacation, retirement, health/life insurance benefits, etc.) the effective salary is approximately $40,000/year. This means that for every hour of disruption, $19.70 is lost in salary-related productivity. Using our office move example, $158/year is lost, for a cost between $1.00 and $1.50/sq ft. To put this number in perspective, the combined annual building utility costs for lighting, power, and HVAC are typically between $0.80 and $1.20/sq ft. Hence, management judgment to move employees one time could have greater impact on government expenses than shutting down the prorated share of the building's total consumption of energy. As this cost impact applies to all downtime occurrences, it is clear that issues of reliability and flexibility must be addressed in building system design to assure that lost time is kept to the absolute minimum.

With the advent of the microcomputer, organizations are now expecting (and usually getting) dramatic efficiency savings in product delivery. Savings of 30 to 70% are claimed from numerous private sector organizations that have monitored production. If these claims can be judged reasonable, even at the lowest projections of improvement, the annual savings per occupant are astounding. A productivity enhancement of 30% in an employee making $40,700/year means that $12,200/year can be justified to achieve that end. In the purchase of electronics support equipment (allowing for a 3-year service life/analysis period and a 10% discount factor), a capitalization of $16,240 would be cost-effective for such an increase in productivity. This relates to $120/sq ft. As a point of comparison, entire building construction costs typically range from $75 to 120/sq ft.

## CONCLUSIONS

The Advanced-Technology Buildings Program offers a significant opportunity to improve the quality of space provided to the federal government. In so doing, the enhanced building features will cost-effectively address improvements in both building operations and tenant productivity. To realize these benefits, however, requires the commitment of both GSA and tenant agencies to work together in establishing requirements and in assuring feedback as to what works and what does not.

# 13

## CHAPTER THIRTEEN

# CABLE MANAGEMENT IN THE HIGH-TECHNOLOGY WORKPLACE

## *PAUL STANSALL AND MICHAEL BEDFORD*

In the face of proliferating information technology (IT) facilities, managers are increasingly being confronted with symptoms of "cable stress," ranging from congested risers (Figure 13-1) and ad hoc cable trays to hazardous, trailing leads around workplaces (Figure 13-2). At the same time, manufacturers of partitions, floors and ceiling elements, and office furniture are making claims for cable management that are often hard to evaluate until the systems have been installed, by which time the money has been spent.

Problems of cable management must be treated in a systematic and interrelated way if expensive errors are to be avoided. This applies to both the design of new facilities for specific targeted clients or speculative leasing and the retrofit of older buildings that were never designed to accommodate IT and its cables in the first place (see Levy's Chapter 15 in this volume). Clearly, the high-technology workplace must be capable of coping with cables efficiently, economically, and in an aesthetically acceptable

This chapter has been reprinted with permission. It originally appeared in *Facilities*, vol. 3, no. 6, June 1985.

**Figure 13-1.** Congested risers—data cable in particular has presented risers with a problem of capacity they were rarely designed to meet.

way if it is to meet the requirements of users. It must also be capable in the future of adapting to unforeseen demands in communications and power connections.

In this chapter, we will first examine the causes of the cable management problem, and refer to economic considerations as well as relevant societal changes. We will then consider the design issues at stake, and examine their implications for future designs. A number of generic design solutions will then be reviewed, with appropriate questions that must be answered before cable management problems can be tackled at three basic levels. Finally, we will present a number of generic solution types that may be used for integrated design solutions at the same levels. To conclude, we will briefly look at the aesthetic impact of properly managed cables.

## CAUSES OF THE PROBLEM

The question that worries most people is whether the problem is going to get worse. Will everyday life in office buildings become intolerable? Will the present systems of cable management become unworkable? Will they become subject to something like the second law of thermodynamics, by which closed systems tend to total disorder or maximum entropy? The an-

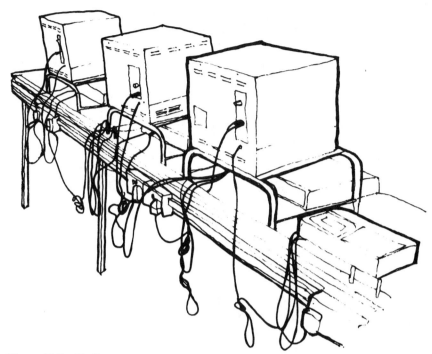

**Figure 13-2.** Trailing leads from outlet boxes to individual places of equipment—hazard and eyesore.

swer to all these questions may depend on how we design and control future cable distribution systems in office buildings. But where do we start? How can we describe the problems in a way that will help to clarify the set of possible solutions?

One line of inquiry could start with the direction in which office buildings are evolving. Many factors are leading to greater differentiation in the office as a building type, which is worrying developers and architects, as well as potential users: outlining the facilities program for a speculative office building is becoming a risky exercise. At the same time the demands from office tenants are becoming highly specialized. Take, for example, the organization that is planning to split itself between its "front" and "back" office functions, ostensibly to reduce the rental overhead in costly inner city locations. The policy-making activities of top management will demand an infrastructure of information services substantially different from the more routine tasks necessary for the back office locations. Top management may depend much more on a reliable system of transmitting information—extending from data and speech to video signals—than on systems of information processing provided by central processors connected to "dumb" terminals. So while a front office may benefit from a building that has a sophisticated infrastructure of telecommunications already built in—for example, block wiring to a central telecom switch for voice and data—a back office might place greater value on the flexibility offered by a raised floor throughout and an ample margin of cooling capacity in the heating, ventilating, and air-conditioning (HVAC) system. Certain aspects of the service infrastructure are more valuable to some tenants than to others. The question is what to provide at the outset over and above the basic building shell.

## ECONOMIC AND SOCIETAL CONSIDERATIONS

Two further trends in the design and use of offices are making us rethink our old ideas about speculative buildings; both are discussed at greater lengths by Eakin and Stitt in this volume. The first is the idea of the "intelligent" or the "responsive" building, which may have a network of sensors for the air-conditioning, heating, lighting, fire protection, and security control so that information on these aspects of the interior environment can be fed into a computer that will either display this to the building staff or use the information itself to adjust the building controls automatically.

The second is the "shared services" building, in which tenants have access to and split the running costs of support services. These may include a host of items such as a private branch exchange (PBX) telephone system, local area networks (LAN), a reception and message service, telex and facsimile transmission, centralized word processing, computer processing, photocopying and reprographics, as well as short-term rental of conference rooms and equipment and the services of a centralized building management staff.

Behind both trends is a concern with running costs. In the intelligent building the concern is to a large extent with energy costs; in the shared services building the concern is with overheads in equipment and administration. Both trends are committed to a high investment in information technology and because of this require careful consideration in the design of cable distribution systems during their inception.

As well as these trends, which appear to be driven by economic considerations, other changes are taking place within the superstructure of society that will have direct effects on the design of office buildings in the near future. The deregulation of the stock exchange in the United Kingdom is a case in point. It signals the emergence of dealing rooms within the investment branches of many large organizations. A stockbroker's dealing room requires a huge investment in electronic communications that are essential in maintaining instantaneous transactions across space.

# DESIGN ISSUES

If these trends are signaling increasing differentiation in office requirements, it is legitimate to consider what design issues they throw up for architects and developers and what office tenants should be on the lookout for when searching for new facilities. Listed below are some of the causes often associated with cable distribution problems, the penalties of which may be increasing building obsolescence:

- Inadequate riser space, poorly located

- Inadequate space for bending radii of cables

- Poorly sited distribution boards and patch panels

- Inadequate storey height for suspended flooring or overhead servicing

- Inadequate space in ceiling voids due to downstand beams

- Insufficient space in floor ducting for cable crossovers

- Inadequate separation of power, data, and telecom cables

- Inaccessible cable trays above ceilings

- Hidden cables running through partitions

- Poorly planned system of outlets at workplaces

- Trailing leads from outlet points to individual pieces of equipment

- Trailing leads between items of equipment

- Underprovision of socket outlets for power

- Disconnected and redundant cables left in risers and ducts

# IMPLICATIONS FOR FUTURE DESIGNS

We tend to think about building design in two ways. First, as a kit of parts that can be replaced over time (shell, services, scenery, and sets); secondly, as a series of drawn projections (plans, sections, and three-dimensional sketches).

## The Kit-of-Parts Approach

The kit-of-parts analysis allows us to identify the critical areas of design in relation to cable distribution. The classification into shell, services, scenery, and sets allows us to consider different rates of expected life and obsolescence of the different parts that come together to provide office space.

## SHELL

- Slab to soffit storey heights
- Size of downstand beams
- Size, number, and location of risers
- Space for distribution boards, junction boxes, and patch panels
- Space for bending radii of cables

## MECHANICAL AND ELECTRICAL SERVICES

- Location of incoming services
- Size and location of ducting runs
- Number and location of outlets

## SCENERY

- Clear height between false ceiling and structure of floor slab above
- Clear height under raised access floor
- Location of partitions in relation to service outlets

## SETS

- Location of furniture and equipment in relation to service outlets
- Size and location of cableways in furniture and moveable screens

One advantage of taking this view is that it helps to clarify who might pay for what in the construction and fitting out of office buildings. While the developer will pay for the construction of the shell and most of the servicing, the tenant would normally pay for the furniture and equipment that go to make up the sets. The issue may be about which elements of the scenery—particularly the floor and ceiling components—the developer would wish to pay for in order to offer value for money to prospective tenants. It is salutary to note that some reported successes in speculatively designed buildings have recently come from developers who have provided flexible ceiling services coupled with raised access floors.

# The Traditional Approach

The kit-of-parts approach is necessary for thinking through cabling issues but needs to be linked to thinking about buildings in traditional terms— plan, section, and three-dimensional views. Cable management problems are experienced within the building interior, where they are easily visible, but they also extend into the building shell, where they tend to be hidden from view and consequently are often forgotten. Because we experience these problems in a disparate way they tend to be treated disparately. What clearly is required is some way of treating and organizing cabling systems in a systematic and interrelated way. This requires us to think of three distinct levels at which cables are distributed around buildings, as illustrated in Figures 13-3 and 13-4.

### PRIMARY
Think of the building in plan: the (generally) vertical distribution of cables through risers and service core ducts from main incoming supply points or from on-floor equipment to each floor of a building (Figure 13-5).

### SECONDARY
Think of the building in section: the horizontal distribution throughout any particular floor—that is, in the plane of the floor, ceiling, or wall.

### TERTIARY
Think of the building interior in three dimensions: the level at which cables emerge from the building scenery elements—that is, ceiling, floor, or wall—and are distributed to and between items of equipment.

The difference between each level is determined by the number of degrees of freedom each one has. The first is constrained within one spatial dimension—a building riser. The second is constrained in two dimensions—through the plane of a floor, ceiling, or wall (imagine it rotating about the point where it emerges out of the riser). The third is free to move in the three dimensions of space. These differences present quite distinct design and management problems. As we move from the primary toward the tertiary level the solution strategies increase while the cost and disruption in rectifying design shortcomings decrease: it is less disruptive to the office to add new furniture than to increase the capacity of risers or extend the underflooring trunking.

Most important of all is the recognition that any satisfactory design should confront all levels simultaneously in order to achieve an integrated design solution. We cannot substitute good performance at one level for lack of capacity at another. Therefore at the facility programming stage of

- ■ **PRIMARY**       I
- ■ **INTERFACE**
  **PRIMARY/SECONDARY**    I:2
- ■ **SECONDARY**      2
- ■ **INTERFACE**
  **SECONDARY/TERTIARY**    2:3
- ■ **TERTIARY**      3

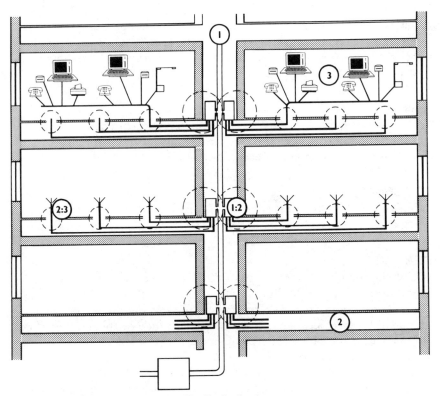

**Figure 13-3.** The three levels of cable distribution.

a project these issues should be given strategic value. At the primary level the problem concerns specifying the capacity, number, and location of ducts in relation to the plan area. At the secondary level, often thought to be the key to cable distribution, there is a wider range of possibilities all of which have implications for the building section.

Just as important as the distinction between the three levels are the interfaces that form the linkages between them. Floor distribution boards, junction boxes, and data panels form the interface between primary and

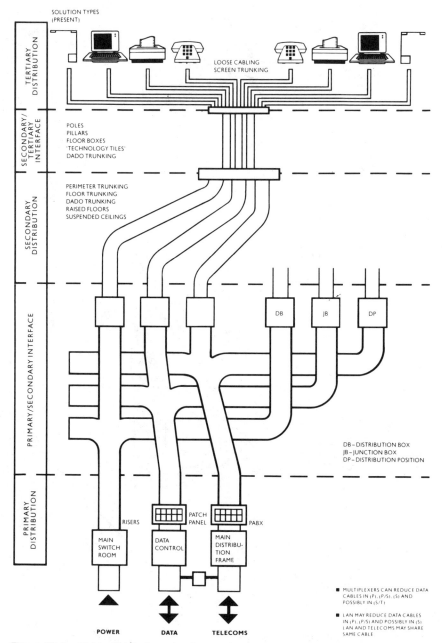

**Figure 13-4.** Cable distribution—levels, interfaces, and equipment.

271

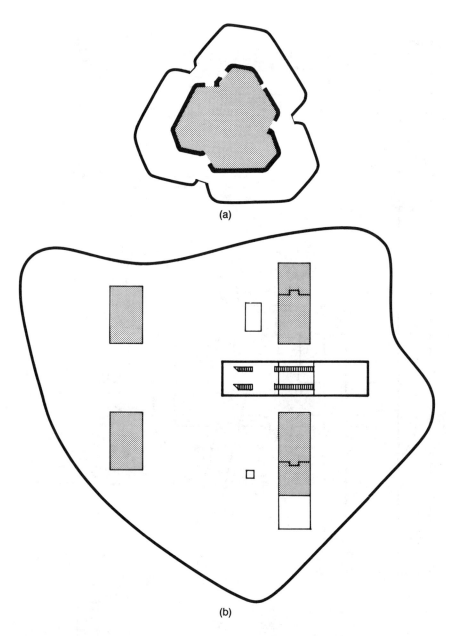

(a)

(b)

**Figure 13-5.** Generic variations on the density of primary distribution points:
(a) National Westminster Bank Headquarters, London—centralized internal core:
narrow-depth office space. (b) Willis Faber Dumas Headquarters, Ipswich,
U.K.—dispersed internal cores: deep office space. (c) Lloyd's Corporation,
London—distributed perimeter cores: medium-depth office space. (d) Wiggins Teape
building, U.K.—dispersed perimeter and internal cores: medium-depth office space.

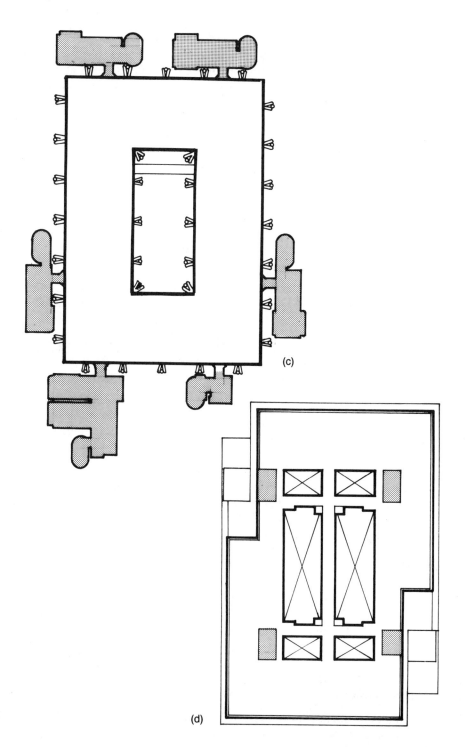

(c)

(d)

273

secondary levels while several solutions are available at the secondary/ tertiary interface ranging from ceiling drops to floor boxes. It may now be useful to consider the classification and evaluation of generic design solutions at the three levels of interfaces.

# GENERIC DESIGN SOLUTIONS

## General Considerations Affecting the Planning of Cable Distribution

- Cost of initial installation and cost of maintenance and adaptability. Flexibility incurs cost premium—how much is needed? How disruptive will maintenance and adaptation be to daily routine?

- Basic capacity throughout the distribution system and allowance for growth (in terms of IT penetration). How mature is the organization? Cable quantity will change in some parts of the system (for instance, LAN will stabilize data cabling in primary and secondary systems).

- Constraints imposed by building shell. For example, what options does floor-to-floor height allow for secondary distribution? What is the impact of structure on distribution options?

- Relationship to other servicing systems. What degree of integration is desirable? Can distribution systems that are expensive for cables alone be justified on the grounds of combination with, say, air distribution?

- Types of layout to be accommodated and degree of flexibility. Can block wiring be considered for all cable services: power, data, and telecommunications? What style is appropriate?

## Considerations Level by Level

### PRIMARY DISTRIBUTION AND INTERFACES WITH SECONDARY LEVEL

- Capacity and distribution of existing risers. There is a trend toward distributed cores—that is, around the perimeter—such as the new Lloyd's Corporation building in London (see Figure 13-5).

- Floor construction. Can additional risers be cut through existing floors?

- Bending radii of main cables. Is there sufficient space for required bending radii of coaxial and other cables?

- Wiring closets. Has allowance been made outside of main risers for distribution boards, junction boxes, and patch panels?

## SECONDARY DISTRIBUTION

- Building section constraints. What options are available at ceiling, floor, or intermediate level?

- Capacity. Is capacity requirement localized or is an even capacity needed throughout any floor?

- Split or single system. Is it sensible to split different types of cable or is there scope for using a single system? For example, would uplighting mean that all power cables could be located at floor level?

- Density of outlets. Different systems offer different densities of outlets.

- Deep plan/shallow plan. Perimeter systems alone are not appropriate for deep plan.

- What structural or other obstacles does the cable system have to work around?

- Can a gradualist approach be employed putting in what is needed as it is needed?

- Safety. For example, is water supplied to the space?

## SECONDARY/TERTIARY INTERFACES

The distinction between secondary and tertiary distribution is that between main horizontal cable distribution about a floor and distribution at and between workplaces. Traditionally these interfaces occur at changes in levels—floor, ceiling, and workplace. With the increasing quantity of cables at the workplace it has been realized that old loose cable connections are inadequate both from a safety and a visual point of view. This has led to a range of ad hoc solutions that manage the interface between the two levels (in some cases, typically shallow enclosed spaces, it is possible to devise both types of distribution at the same vertical level—"dado" height—in which case the interface is simply the junction between secondary and tertiary trunking). These elements are becoming important interior design features, and some secondary distribution manufacturers keen to offer an integrated approach are addressing the issue. Their implications extend beyond basic issues of safety and capacity and relevant criteria are the following:

- Degree of fixity of furniture layout (determines degrees of freedom needed in interface design)

- Type of secondary distribution (solution types are sometimes unique to a particular system)

- Approach to visibility of cable services

- Integration with other services (notably lighting)

- Compatibility with furniture and scenery elements

- Vulnerability (are outlets concealed by furniture?)

- Type of layout (individual offices are a different problem from open plan)

### TERTIARY DISTRIBUTION

Options for tertiary distribution are more limited, and for many low-density users of IT simple loose cables remain viable. The relevant criteria are:

- Amount of kit at the workplace—some office users have a PC workstation with keyboard, video display unit, disk drive and printer, task light, calculator, telephone, and even a microfiche reader, most of which require separate power supply and outlets.

- Rate of change—how accessible do cables need to be at the workplace?

- Compatibility with existing furniture—do furniture suppliers offer cable managers as an add-on?

## GENERIC SOLUTION TYPES

### PRIMARY DISTRIBUTION

- Risers and wiring closets

### SECONDARY DISTRIBUTION (Figure 13-6)

- Flat wiring—under carpet

- Floor trunking—regular grid, perimeter/corridor; main runs with or without spurs

- Suspended booms or trays

- Skirting trunking

- Wall-mounted, dado trunking

- Suspended ceiling—whole or part

- Raised floors—shallow (normally cables only) or deep, high capacity (for example in computer rooms), sometimes with other services

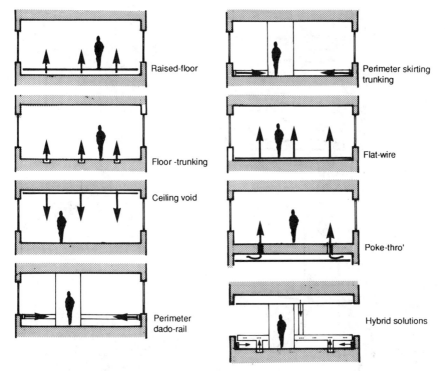

**Figure 13-6.** Generic variations on the organization of secondary distribution.

- Partitions—traditional fixed walls or demountable partitions with cableways

## SECONDARY/TERTIARY INTERFACES (Figure 13-7)

- Poles (with ceiling and/or floor distribution)
- Pillars (floor)
- Flexible cable drops (ceiling)
- Panels—ceiling and/or floor; fin walls, technology panels (in partition system), technology panels (in furniture screens)
- Loose cables—floor boxes
- Loose cables—from fixed outlets
- Dado trunking

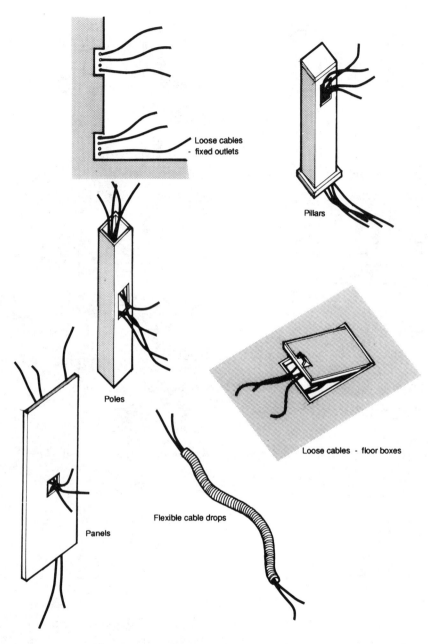

Loose cables - fixed outlets

Pillars

Poles

Loose cables - floor boxes

Flexible cable drops

Panels

**Figure 13-7.** Secondary/tertiary interfaces—good cable management needs a safe and elegant interface between secondary and tertiary distribution. The traditional solutions—loose cables in floor boxes and loose cables in fixed outlets—are neither safe nor elegant.

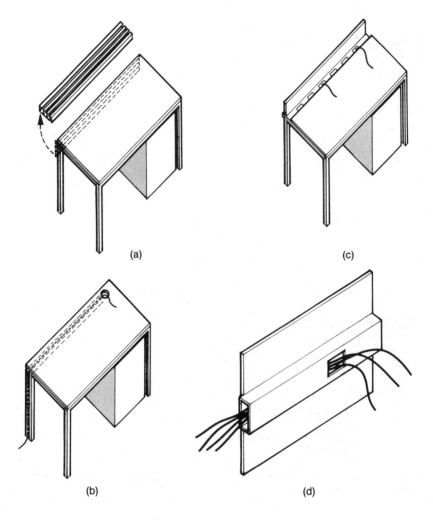

**Figure 13-8.** Tertiary distribution: (a) Clip-on cable managers. (b) Built-in cable capacity. (c) Integrated cable distribution (d) Screen-mounted trunking.

## TERTIARY DISTRIBUTION (Figure 13-8)

- Screen or wall-mounted trunking

- Furniture—clip-on cable managers, built-in cable capacity in structure, integrated cable distribution

- Cable ties—for tidying up loose cables

- Loose cables—in other words, absence of cable management

**Figure 13-9.** The technology panel belies its name, disguising technological elements as interior design solutions.

## ELEMENTS OF STYLE

Increasing use of IT means that integration of the different levels of the cable distribution system is critical for the efficient and safe functioning of the building. The cable demands on the 1980's office also require a policy on the visual impact of cable distribution. The market now offers the possibility of orchestrating a solution at either end of the spectrum. At one end lies tasteful concealment (Figure 13-9)—a fully accessible raised floor into the outlets of which are plugged furniture using a range of flaps and hidden conduits. At the other is the high-tech, expressive solution (Figure 13-10)—suspended, open mesh cable tray with flexible hose cable drops plugging into screens and furniture. Lack of consideration for image makes many offices a disappointing mixture laying somewhere between these poles.

## CONCLUSION

Although the causes of cable management problems are manifold and may be experienced disparately, they can be treated systematically. It does however require us to seek design solutions that integrate the various levels and interfaces described here and that are commensurate with both capital and operating costs and thereby related to both developers' and users' expectations. By including cable management issues at the earliest stage in the facilities program then the design team is more likely to approach them strategically and work together for an integrated solution. It is no longer practical to rely solely on either interior specialists or product suppliers in the expectation that a complete cable management strategy will emerge. The price of expensive partial solutions in the long run may lead only to increasing building obsolescence.

**Figure 13-10.** The high-tech solution is a legitimate response—but unless carefully considered, its expressive image may be difficult to disentangle from an absence of planning.

CHAPTER FOURTEEN

# INTELLIGENT BUILDINGS, TODAY AND IN THE FUTURE

## FRED A. STITT

The ubiquitous information revolution has now spread into the very fabric of our buildings. The human-made environments that shelter our daily activities, are no longer the passive shells they have been from the beginnings of civilization. They can be active and respond to changes in external constraints and user requirements. The following are predictions by consultants in the field of *intelligent buildings.*

• From now on (1987) virtually all new buildings of substantial size will be designed as "intelligent."

• During the next 10 to 15 years all existing buildings of substantial size will be retrofitted for "intelligence."

• By the end of this century all buildings of all sizes and ages will have been fitted, and probably refitted, with the latest level of "smartness."

The advent of "intelligence" as an integral part of buildings is a recent development with far-reaching consequences and implications. This chapter discusses *intelligent buildings* (IB) as the merger of information and telecommunication technologies, and construction technology. We will first present what we mean by intelligent buildings today, and then consider what they may become in the short and long terms. Finally the reader will be invited to speculate about some future possibilities, from the springboard of current cutting edge research, and to appreciate their implications for human intelligence.

## INTELLIGENT BUILDINGS TODAY

Before we deal with what "intelligence" means in this context, here are some further comments and speculations from IB consultants:

- Buildings owners may earn more from electronic services they provide to tenants than from space rentals.

- Intelligent buildings might cost an additional 1% in final construction, but may have a payback period of only 2 years.

- Intelligent buildings best meet the needs of high-tech tenants, and the relationship will spawn new varieties of building enhancements as yet barely dreamed of.

Any market that will encompass *all* new and *most* old construction should attract vast amounts of corporate resources and capital, and it is doing so. There are billions of dollars at stake, and so IBM, AT&T, Honeywell, GTE, Mark Controls, United Technologies, and hundreds of smaller IB-related companies have swarmed into the field.

What is an intelligent building compared with a dumb one? There are two areas of possible differences:

1. Internal controls for automated management of building equipment. Sometimes called *centralized building systems management.*

2. Varied *shared tenant services* (STS in the jargon of the trade). STS tenants rent communications or computer services including discount long-distance phone services, computer workstation rentals, and software, and pay only according to usage.

Some technological aspects of both areas are discussed in Chapter 12 of this volume by David Eakin in his review of advanced-technology buildings.

In the building control area, there are numerous tenant conveniences and owner economies such as the following:

- High-response elevator controls that detect movements of people before they get to the elevators.

- Fire controls that help with smoke removal and fire-fighting management.

- Individualized lighting and heating, ventilating, and air-conditioning controls that turn lights and climate controls off at specified times after everyone has left a space.

- Building equipment that monitors itself and telephones for help when there is a problem.

- Security systems that sense potentially harmful human presence and movement.

Automatic and highly sensitive building controls are one major characteristic of intelligent buildings. The other, shared tenant services, can include the following:

- Lowest-cost long-distance telephone routing

- Call forwarding

- Message centers

- Access to computer service bureau main frames

- Workstation and peripheral hardware rentals

- Software and data base library access

- Facsimile transmission

- Teleconferencing

- Reprographics and copier services

- Automated secretarial, dictation, and word-processing services

- Telex

- Electronic mail

- Cable and satellite video

- Data base library access, information, and research services

All these services can be offered to tenants at low prices because of the bulk buying power of the STS landlord. Other advantages to tenants are the following:

- They avoid capital expenditures.

- They avoid problems related to 2- to 3-year obsolescence of new equipment.

• Through well-planned "prewiring" they do not face major expenses every time they move a computer terminal (up to $1,000 per terminal in many cases).

The two essential divisions of intelligent buildings—centralized building systems management and shared tenant services—have much in common, but they are not necessarily integrated as one total system. They are analogous to the two main separate functions of the human nervous system, one being automatic and dedicated to internal body functions and maintenance while the other responds to needs and changes generated by external circumstances. There are similarities between the systems, and both use many of the same sensory and communications apparatus, but they remain functionally distinct.

The sensory and communications apparatus of an intelligent building start with a central wiring (or fiber-optic) communications spine that runs vertically through the building. The spine or communications core is often called a *data highway,* and it is separate from traditional electrical power wiring. Microprocessors are hooked into the data highway for sensing change in the building's systems and modifying system controls. Heat and cold, air pressure, vertical transportation, lighting, and detection devices are all monitored and modified according to need. Action devices such as alarms can trigger intercom warnings and telephone calls for outside assistance. Here are some specific functions that can be handled by the data highway, microprocessors, and sensory devices:

• External temperatures are monitored to forewarn the central mechanical system of upcoming needs.

• Morning arrival of occupants is detected, which gets the elevator system geared up and activates heating or cooling systems.

• When work spaces are entered, infrared detectors switch on lights as needed.

• If there is a malfunction, such as an elevator shutting down, the elevator system will call for help and inform the elevator's occupants as to what is happening.

• If there is a true emergency such as fire, alarms and synthesized voice warnings will give instructions to occupants; fire fighters will be called and directed to the fire area; elevators will automatically go to required locations; and the central air-conditioning system will pressurize spaces around the fire and exhaust smoke from the hot spots.

During the day, building services will respond in numerous major and minor ways to tenant and building needs:

- When occupants of an office leave in a group to head for the elevators, corridor sensors will alert the elevator command center of the traffic that is coming.

- If elevators or corridor sensors detect fast, possibly violent human movements, a call will go out to security staff. Security will also be called whenever human presence is detected in off-limits or inappropriate areas, such as when someone is loitering in a stair well.

- When a room empties out, the lighting shuts off after a predetermined number of minutes. The heating or air-conditioning begins shutting down too unless the system knows by timing or programmed instructions that the departures are only temporary.

Imagine the possibilities for high-tech buildings if you extend and amplify the sensory and action features just described. Exceptionally precise controls of temperature and humidity, specialized lighting, microscopic decontamination of air, differential air pressure controls, security systems, and spotting building system failures before they get out of hand—all these can be routine aspects of the design of intelligent buildings. The basic wiring and control systems are well understood, and so it is just a matter of custom selecting sensory and response mechanisms required for special high-technology environments. Add in the ultrahigh level of communications and computer needs of high-technology tenants, and it is clear that the IB concept and high-tech buildings amount to one of the world's rare ideal marriages.

A fortuitous combination of events made all these developments possible, and they point to what we can expect in the future. A key component of shared tenant services, for example, is the right to operate a private telephone branch exchange (PBX). In the United States the deregulation of the telephone industry made that possible, and without shared telecommunications there would be no total package of shared tenant services to offer. Further deregulation throughout all areas of communication is still far more likely than a regression to government controls. (The exception is in state law. Some states are blocking advances in communications with restrictive legislation.)

Deregulation has other related effects. It is creating explosive growth in new communications, shipping, transportation, investment, and service companies. These new wave companies are among the leading prospective tenants for the new buildings.

Other historic combinations of events that culminate in intelligent buildings include the improvements in sensory devices and control systems, especially as developed in aerospace industries; computerization in general; and, of course, the relentless development of the chips and microprocessors that in their operations allow intelligence to be distributed, multiple, and highly adaptive instead of centralized and nonadaptive. This historic trend, our current Industrial Revolution, as yet has no perceivable limits.

A social/business trend that matches the preceding ones is the growth of small technologically oriented businesses. Business writers have observed for years that most new jobs are created by the new small entrepreneurial companies, not the old ones. They need access to large electronic data-processing and telecommunications resources on a shared pay-as-you-go basis, and they are a major and growing market as potential tenants for the smart buildings.

Smart houses will be a parallel development to the smart working environments, and many people will choose to establish extensive communications and work links between home and office. Some housing developers are already prewiring houses for extensive communications functions and internal systems monitoring.

As a design professional, building owner, or developer, how does one get access to information about all this? Although the subject is new, there is a small information explosion in process. There is an increasing availability of consulting services, telecommunications companies, full-service companies (such as AT&T, Honeywell, IBM, and United Technologies) that will be glad to respond to information requests. One of the best ongoing sources of data in this field is the "Architectural Technology" portion of *Architecture Magazine*, published by the American Institute of Architects.

Meanwhile, and partly because of the very newness of IB, there will be controversies as to which cable systems are the best; how much space you need for wiring and for peripherals to smart systems; what kinds of floor or ceiling systems work best; how much electrical power you need for the next generations of office equipment; how best to plan, cost estimate, and get aids for your communications needs; how to handle retrofits; whether to use design professionals, engineers, telecommunications consultants, or one-stop service companies in putting together a system; whether systems are best handled by building owners or independent companies.

All of the above are already the subjects of widely divergent opinion, and a few formerly up-and-coming companies in industry have already folded. So, promising as all this is, it is also a volatile business that deserves considerable homework before major decisions are reached. The appended bibliography is intended to help the reader in this homework.

## INTELLIGENT BUILDINGS: WHAT THE FUTURE HOLDS

### The Short Term

In the short history of avowedly intelligent building design several divergent trends have appeared. One is the growing decentralization of the buildings' brains and nervous systems. Whereas the first room sensor light switches were controlled by a master computer, for example, now autonomous switches are available that do effectively the same thing: turn on lights when someone enters a space, and shut them off a few minutes after the room is empty.

Another trend is the further microcomputerization of buildings, which we can expect to see in much the same way that we have seen widespread microcomputerization in business and industry. Smaller areas of buildings will be monitored and controlled independently, as will individual rooms. Where it is important for building management to record what is going on overall, the small computers will provide reports through networking. Small computers will be in charge of the controls for individual areas instead of master mainframes, as was the case in the past.

How far will miniaturization of intelligence go? Ultimately, down to the molecular level. Between now and the 1990s we will see periodic quantum leaps in miniaturization, cost reduction, and versatility in building controls—all comparable with what we have already seen these past 5 years in the world of business microcomputers.

Along with decentralization comes a related development in intelligent buildings, which is unfolding for all computers—namely, *artificial intelligence* (AI). This term refers to intelligent building facilities that combine sophisticated sensory and response apparatus with *expert systems* or similar intelligence mimics that analyze problems and test out solutions.

Allied to artificial intelligence and expert systems are spoken language functions—both as input and output. Today's rather dull talking elevator gradually will become an information service for the riders that is capable of synthesizing speech to report on needs for repair work that it cannot manage on its own. *All* other major building appurtenances will take on similar communication roles. None of this should be mistaken as futuristic speculation for it is just the inevitable continuation of well-established technology and trends.

Building repair, maintenance, and security can and will also be robotized while being enhanced with AI and verbal skills. Semiautonomous office delivery, floor cleaning, fire hose carrying, and police bomb suppression robots are already on the job. And although U.S. industry has been slow in robot development, the Defense Department is racing the Japanese to see who can produce the greatest robot breakthroughs in the next few years. Robotics will require catch-up study by building planners. They have to become alert to the need for generous prewiring for added communications and power circuits needed by the wheeled servants that are coming on the market.

Futuristic expectations are sometimes so many years ahead of the current reality that by the time the predictions come true, they seem like old stuff. That will be the case with robots. They are merely mobile, programmable tools, and the early versions won't seem like much more than roving appliances. But a few years down the road we will see power and versatility increase a 100-fold—another rerun of the microcomputer story—until the robots, like our buildings, combine such diverse conversational and analytic skills that they will seem more like companions than machines.

Along the way, other technological evolutions and revolutions will grow, extend themselves, merge with one another, and create whole new worlds of opportunity and problems for facilities planners and managers. For example, as all media become digitized, processes that used to be as different as photography and video, photocopying and computer graphics, offset printing and telephone communications, movies and laser disks, will all merge into one multifaceted electronic medium.

Still or moving pictures will be recorded electronically (with sound, if wanted). They will be transmitted from "camera" to phone, transmitted by modem to computer, shown on video, and "played back" in full color and fidelity on office copiers. Combine these capabilities with building and robot optical and audio sensors and the world of information processing enlarges accordingly. In other words, it won't be only people creating, recording, transforming, and transmitting information. Our tools and living and work habitats will do the same in the course of doing their jobs.

## The Long Term

Enter the Dynabook and Xanadu. These are "old" future concepts that are gradually coming into reality. The *Dynabook* is a proposed information storage package possibly the size of a lap-size computer that will simply hold all the information there is. Anything you want to know about or refer to or get a print of will be there in the Dynabook. All the hundreds of diverse information services that are growing up around the world are just small parts of the ultimate information device. The Dynabook is the dreamchild of Xerox pioneer Alan Kay. It and similar devices by other Silicone Valley theorists are contemplated to ultimately not be "books." Rather they will be invisible, just built into the environment like so much wall fabric.

*Xanadu* is the name of another bold informational goal: mainly the categorization, and hence *pre*creation, of all possible categories of information that can possibly exist. One objective of Xanadu is to create an electronic international publishing system whereby anyone can input any type of prose or poetry, speculation or research documentation, design or drama, and be paid for the contribution to the massive data bank in proportion to the amount of time other people hook in and scan or copy their inputs.

Thus original new art, literature, music, news, political commentary, cartoons, problem/solution analysis, ideology, theology, education, games, all enter the massive special interest electronic flea market. Everyone's work is accepted or rejected not by go-betweens and arbiters of taste such as editors and publishers, gallery owners, impresarios, or faculty committees. Instead success or failure is determined strictly by the customers.

Imagine exchange between independent contractors offering such professional services as electronic medical exams and diagnosis, legal advice, and design services; or between workers who input data, make phone calls, sell products, and do bookkeeping: or between technical employees who do engineering calculations, write software, and edit manuscripts. As many people already do, they will be able to work anywhere and for anyone. They will not have to be in office skyscrapers or peripheral office/industrial parks.

What does this have to do with the intelligent buildings? It means that intelligent or not, if they do not adapt to the dominant work/social change of our time, the buildings will wither and die of high vacancy rates.

There is another technological link that will upset the traditional relationship of building and tenant. Along with the miniaturization and increased computer power come increasingly sophisticated image processing, then video/computer interface, and now *simulated reality* or *artificial reality* (AR). Ray tracing, fractal algorithms, and the like are creating remarkably realistic images that already promise to transform the moviemaking industry.

More significant than simulated reality from a design standpoint is *recreated reality*, that is, images of that which does not or cannot exist. One such image, the fourth-dimensional cube of Tesseract, can now be seen in all its three-dimensional glory as a hologram that "impossibly" turns in and out of itself as it revolves.

All art and design media can become environmental media. The functioning *intelligent environment*—particularly the home environment—will create whatever images are desired as well as whatever images and information one wants to have transmitted for whatever purposes. Thus can we be surrounded by environments of transcendent beauty and scope or unspeakable banality, depending on the tastes of the user. Then the intelligent building becomes part of a larger communication medium and process: ourselves, all of us, dealing with the innermost workings of our own psyches.

## Speculations on Future Directions: a Conclusion?

All this is no more than extrapolation on what already exists and/or is in process. It sets targets for current research and has been well worked out in the research centers of the Massachusetts Institute of Technology, Xerox, IBM, Apple, Bell Labs, and the like. So, if this describes the not-too-distant future, what will follow after these stages? There are two special developments to plan for. One bears greatly on practical issues of how we design our intelligent environments for human use. And the other allows us to enter a realm of speculative invention that is about as far-reaching as we can discuss with words and concepts currently available.

The first of these contemporary trends with extraordinarily vast potential consequences is a natural outgrowth of artificial intelligence research. That is the realm of cognitive sciences or "applied epistemology"—the study of our nervous system, consciousness, and the nature of knowledge.

A rudimentary survey of human nervous tissue shows we function with a staggeringly high level of cognitive potential. Our brain, only one part of the massive intelligence that pervades our bodies, routinely processes something like one quadrillion bits of data per second. A rough analogy of that volume of data could be presented by running average English words across a block-long video screen at a rate of 15 billion words/sec. (A comparably dramatic image of brain functioning is provided by the estimate that the total number of switch interconnections in the brain is larger than the number of electronic particles in the observable universe.)

It is exhilarating to contemplate the enormity of this intellectual potential. But exhilaration is followed by depression as we also realize that we are not running our nervous system anywhere near its potential. Even those who think they are racing right along are barely on idle. Finally, new excitement sets in once it is realized we have one of the great mysteries of all time to solve and some beginnings of the understanding needed to solve it.

The mystery leads inevitably to the microworld of our own sensory/perceptual/cognitive hardware—to the subatomic particles of which we are made and the system of construction that brings forth protein, animation, and intelligence. Chances are that the best way to understand it is to duplicate, manufacture, or grow our own stuff of life and intelligence.

Thus enters the other most potent direction of new research of all time—miniaturization and manufacture of tools, machinery, and computers right down to the molecular level.

Eric Drexler, author of the 1986 book *Engines of Creation* and recently of the Space Sciences Lab at MIT, is one of the pioneers in the realm of molecular scale design—that is, he designs hypothetical machine and computer components that are the size of molecules. Although some people doubt that design and manufacture on that level would be possible, Professor Drexler can respond with numerous examples, already fully functioning in the world of viruses, crystal growth, and biochemistry. They already exist; we have only to duplicate them and put them to work.

I asked Professor Drexler not long ago about the implications of his similar research on materials, environmental design, and buildings. He pointed out applications for which the term *intelligent building* will be a much more accurate description than it is for any buildings we have now. Keep in mind that he envisions molecular-size computers with megabyte random-access memory acting as parallel processors and being self-directed components of larger objects. Any object you can imagine could "come together" as molecule-size components assembling themselves and taking on specified characteristics. In the same way, they could disassemble: a wall, a machine, a printed page, a sculpture—even a hot meal with a glass of wine—all solid, real, intelligent, and "artificial."

Structural materials, for example, could be designed with billions of built-in molecule-size parallel processor components. Thus structure could respond to changes in external forces, redistribute itself to match changing conditions, repair itself, and most importantly, provide efficient support with a miniscule percentage of the mass that is presently required in building construction. Mechanical systems can be similarly intelligent, responsive, and self-adapting. Electrical systems can also, and so can every square inch of the environment and its appurtenances.

How will we design on this level? No one knows yet, although musical and organic structures offer some starter examples. Meanwhile, Professor Drexler suggests some other avenues of opportunities for his molecular machines. They could travel and do tasks. They could enter the body and the nervous system, do repair work, clean out damaged DNA in the cells, reattach damaged nervous tissue, and bring damaged human intelligence and consciousness back to health and focus.

This "second-generation" speculation beyond what is immediately possible pulls us back to first principles and considerations. It pulls us back to take a closer look at what it is that we are designing buildings for in the first place, whether they be dumb buildings or smart ones. Presumably we make intelligent buildings for the purpose of serving the needs of intelligent creatures—ourselves, of course. And the highest level of service to intelligent creatures is to help them make themselves more intelligent.

On a functional level we achieve that by creating environments and tools that reduce the laborious and repetitive aspects of life and work. And we plan environments and tools so as to minimize stress and frustration. In so doing we open up the opportunity of more time for people to spend in learning, directed problem solving, introspection, creative speculation, etc.

Moving beyond the functional, we can also enhance the sensorial, perceptual, and cognitive environment by providing aesthetic pleasure. It is observable that people literally "open up" when they experience pleasure. They open their receptors, their perceptual channels, their elecrochemical emotional states, and their willingness to continue experience and be conscious of *more*—more in the environment and more internal experience.

Aesthetic response is usually a mixture of sensory, perceptual, cognitive, and emotional pleasures—a larger mix than is provided by most any other kind of experience. Since architecture and environmental design might be thought of as the largest, most engulfing potential source of such experience, perhaps we should create and evaluate architecture and environmental design from that standpoint.

The primary standard of value of what building, architecture, and environmental design are all about may well be the mind—the human consciousness. The architecture which enhances and promotes the continued existence and growth of consciousness would be considered successful. And buildings that do not, whether "smart" or "dumb," would not.

## BIBLIOGRAPHY

### Articles

Eley, J. "Intelligent Buildings." *Facilities* 4(4), (April 1986), pp. 4-11
Fisher, Thomas. "Intelligent Architecture" *Progressive Architecture* (May 1984), pp. 167-172.
Gannes, Stuart. "The Bucks in Brainy Buildings." *Fortune* (December 24, 1984), pp. 132-145.
Hall, Gary. "Wired for Change." *Architectural Record* (Spring 1984), pp. 14-24.
*Interior*, April 1985
"Picking the Brain of the Intelligent Building." pp. 13-14.
"Intelligent Building." pp. 101-114.
"Intelligent Buildings: IT into IQ." pp. 116-117.
"Intelligent Buildings: Programming." pp. 118-119.
Thompson, B. "Two 'Intelligent Buildings' Associations Organizing." *Building Design & Construction* (November, 1985), pp. 29-32.

### Periodicals

*Business Communications Review*. 950 York Road, Suite 203, Hinsdale, IL 60521.
*This Month*, The Monthly News Digest of Key Data Communications and Telephony Events. McGraw-Hill Information Systems Co., 1221 Avenue of the Americas, New York, NY 10020.
*Architectural Technology*, AIA Service Corp., 1735 New York Ave. N.W., Washington DC, 20006.

### Books or Manuals

Gouin, M.D., and T.B. Cross, *Intelligent Buildings—Strategies for Technology and Architecture*. Homewood, Illinois: Dow Jones-Irwin, 1986.

Schwanke, D., and K.S. Roark. *Smart Buildings and Technically Enhanced Real Estate.* Urban Land Institute, 1090 Vermont Avenue, NW, Washington, DC 20005, (202) 289-8500.

Sinopoli, James. *A Management Guide to Office Information Networks.* Office Telephone Management Inc., 5524 Bee Cave Road, Austin, TX 78746, (512) 328-3880.

**PART FOUR**

# EXAMPLES OF PROJECTS FOR HIGH-TECHNOLOGY WORKPLACES

# 15

## CHAPTER FIFTEEN

# THE REUSE OF OLD INDUSTRIAL BUILDINGS FOR NEW PRODUCTION PROCESSES

## DAVID LEVY

The high-technology workplaces of the future are likely to be housed in the buildings of the present. The reuse of existing buildings by the industrial, commercial, and institutional sectors of the U.S. economy is currently the predominant approach to procuring space. Figure 15-1 shows that 83% of total industrial, commercial, and institutional construction dollars spent in the United States in 1984 was for reuse projects. Of the $111 billion in total construction in these three sectors, $92.4 billion was spent on reuse projects. The 1985 estimate for reuse construction was $101.6 billion, or 81% of the total construction. Between 1977 and 1985, the real dollars spent on reuse increased from $21.6 billion to $101.6 billion, while spending on new construction declined from $31.4 billion to $23.4 billion.

This chapter presents an alternative view that focuses on how firms can and do reuse old industrial buildings to house new production processes. The factors underlying this shift toward a decided preference for reuse construction are explained. Some of these factors include the society's preservation ethic, government subsidies of business, and changes in the in-

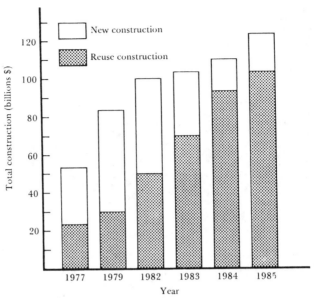

**Figure 15-1** Commercial, industrial, and institutional reuse and new construction expenditures 1977–1985 (source: *Building Design and Construction*, May 1984, p. 21).

dustrial real estate market. Three general strategies for industrial building reuse are described along with the key aspects that determine a project's feasibility. A set of risk factors dealing with costs associated with doing any reuse project are presented. The conclusion of the chapter describes two examples of reuse projects: the renovation and expansion of a small industrial facility for continued industrial use and the conversion of a large vacant industrial plant into a small business incubator.

## INCREASING PROMINENCE OF USE

The reuse of old industrial buildings for new production processes is flourishing in view of changes in how society views these buildings. A preservation ethic now exists in America that was not present in the urban renewal era of the 1960s, an attitude that began with the preservation and reuse of historic buildings and neighborhoods and has now expanded to include a range of building types, including factories.

Old vacant factories are poignant and constant reminders of lost jobs and prosperity. The social and economic costs of a plant closing are staggering for most communities.[1] The costs associated with these closings have been providing a powerful stimulus for politicians to do something with the vacant plants.

Corporations have adopted a more progressive view toward old industrial buildings. Corporations have found that a closed plant is an expensive liability to hold on to and maintain.[2] Holdings costs can run as high as $3/sq ft/yr. These costs, plus a desire to present a socially responsible image, have caused corporations to seek out new ways of disposing of or reusing their surplus properties.[3]

Government incentives have provided a big boost to the reuse of industrial buildings. Politicians know that jobs mean votes, so they promulgate various programs to encourage economic development. Tax abatements for industrial building rehabilitation are a popular state government subsidy for business. Programs developed at the state level have been criticized for actually causing unemployment due to firms' becoming more capital intensive,[4] creating windfalls for larger firms,[5] and increasing the tax burden on individuals.[6] The end result of these competing programs to subsidize business has been to pit the states against one another in an economic development fight.[7]

Improvements in building technology have contributed to the growth of industrial building reuse. Until the late 1960s few building contractors had the skills to do reuse projects. Many contractors have now acquired these skills and specialize in rehabilitation work. Industrial engineers and managers have found new ways to integrate modern production and materials handling equipment into old buildings.[8] Managers find it easier to participate in the planning process for a reuse project because they are familiar with and understand the buildings involved.[9]

Industrial buildings are reused more frequently today because it is profitable to reuse them. Reuse projects can be completed more quickly than new construction projects. Shorter projects mean less production downtime and disruption to production operations. Construction costs for reuse projects are typically less than for new construction because costly foundation and structural work is in place.[10] The costs of moving operations and equipment are low for a reuse project, because moves within a building are easier than a move to another facility.

The opportunities for financing a reuse project have improved. Financial institutions were cautious about financing some of the earlier attempts at reuse because they were not knowledgeable about the problems involved. The level of risk was high because of inexperienced contractors and the unforeseen problems that accompany the renovation of any building.

The demand for small industrial buildings was very strong between 1980 and 1985 because of a rapid growth in new business. While a surplus of very large industrial buildings exists in many parts of the Northeast and industrial Midwest, there is a corresponding scarcity of good-quality small buildings. With the demand for small parcels of space high and the supply low, many communities now have small business incubators occupying large industrial buildings that once stood vacant. The incubator concept has proved to be one of the most successful strategies for the reuse of vacant factories.

## REUSE STRATEGIES

Initially, selecting a general reuse strategy depends on the size of the building in question. As building size increases, the likelihood of reuse by a single tenant decreases. Large buildings over 75,000 sq ft (6,967.5 sq m) are usually older buildings. The demand for large parcels of old space is low because large firms can afford to build what they need, and they will seldom reuse what a similar firm has already vacated. At the other end of the scale, the demand for small parcels in the range of 3,000 to 15,000 sq ft (278.7 to 1393.5 sq m) is high because of the rapid increase in the formation of new businesses. These market factors point toward three different reuse strategies for industrial buildings:

1. Small buildings for small firms

2. Large buildings made small

3. Large buildings for large firms

Buildings under 75,000 sq ft offer the greatest opportunity for reuse by a single firm. In Michigan 65% of firms that recently completed rehabilitation projects needed less than the 75,000 sq ft of space even after expanding by an average of 50%.[11] Small firms can seldom afford the high cost of new construction, so the reuse of an existing building is always a viable alternative.

The second strategy, large buildings made small, is the most effective way to reuse large industrial buildings. Large vacant buildings are often available at very low prices because the demand for them is low. But this may be changing because of the incubator concept. An incubator building provides small parcels of space and business support services like word processing, security, telephone answering, and technical assistance to new small businesses at a total price near the market rate for just the space alone. Such low rental rates are possible because incubator buildings are usually acquired for little money and many receive government subsidies to cover initial start-up costs. There are around 120 incubators currently operating in the United States, with most of them located in the Northeast and Midwest. (See Davidson's Chapter 6 in this volume for a detailed discussion of incubators.)

Large buildings are seldom reused by single firms because of the age and condition of the building, inappropriate location of the building relative to the firm's markets, irregular arrangement of structure and services, and high utility costs. But there are several examples of large firms successfully reusing large, old industrial buildings for new production processes. In the automobile industry the most notable example is the Chrysler Corporation's Jefferson Avenue Assembly Plant in Detroit where the K-car is assembled. The electronics industry has at least four good examples. The Apple Macintosh is assembled in a renovated and highly automated old factory in Fremont, California. GenRad Corporation manufactures electronic testing and analysis systems for the computer industry in a renovated facility in Concord, Massachusetts. In Manchester, New Hampshire, Hendrix Electronics manufactures text-processing equipment for the publishing industry and optical character recognition machines in a building originally constructed in 1838 as a part of the Amogskeag Millyard. General Electric Company manufactures its entire line of electrical metering products for consumer and commercial applications in a large, 65-year-old five-story building on the west bank of the Salmon Falls River in Somersworth, New Hampshire.

## ASPECTS OF FEASIBILITY

In addition to building size, the feasibility of reusing a particular industrial building also depends on the building's overall condition, type of structural system, local building code, site considerations, and special risk factors associated with renovation work. While each situation is different, a set of general principles can be applied when assessing the feasibility of an industrial building for reuse. These feasibility principles are defined in Table 15-1.

The condition of a building's mechanical and structural systems, walls, windows, and roof is the most obvious factor affecting feasibility. Roof problems are the most common and can account for up to 20% of the construction costs for a typical reuse project.[12]

The programmatic needs of a firm must be broadly met by the building that is a candidate for reuse. The building has to be of the appropriate size and configuration in order to support the production needs of the company. Configuration is most important since it determines the possible layout of the production process. A regularly spanned arrangement of structure and mechanical services is a basic requirement for a linear flow of production and materials. So is the ability to keep the flow on one level, unless level changes can occur automatically with such means as conveyors or automatic storage and retrieval systems (AS/RS). If material handling is extensive and the pieces being handled are large, multilevel buildings

---

**TABLE 15-1**

**Key feasibility aspects in the reuse of industrial buildings**

---

**Building Size**
- If the building is less than 75,000 square feet it will be more appropriate for a single tenant.
- If the building is greater than 75,000 square feet it should be considered for multiple tenants or as incubator space.

**General Condition of the Building**
- What is the condition of the structural, mechanical, exterior cladding, and roofing?
- Roof problems are the most common and can account for 20 percent of the construction costs on a reuse project.

**Programmatic Needs of the Occupants**
- Is the building the appropriate size and configuration to accommodate the production requirements of the potential occupants?
- If the building is multi-story can materials be moved efficiently and consistently between floors?
- Does the building's configuration allow for production equipment to be layed out in a linear fashion?

**Regular Arrangement of Structure and Services**
- Are the structure and services organized and arranged in a regular fashion?
- Does the building have numerous, irregular small additions that have resulted in a non-linear building configuration?
- Through alteration and modification, have the services and structure become highly specialized for a limited number of operations?

**Building Codes**
- Local codes may not address the problems encountered in reusing old buildings.
- Building inspectors tend to enforce the code more strictly for rehabilitation projects.
- Code compliance for reuse projects takes longer than for new construction projects.
- Code compliance focuses on occupant safety and typically has a negative impact on the economic feasibility of a project.

**Available Knowledge About the Building**
- There is typically very little technical knowledge specific to a particular old industrial building.
- The building maintenance person can be an invaluable source of information on how the building works.

**Site Considerations**
- The site coverage ratio should not be greater than 50%.
- Is there adequate space to allow large trucks to have access to the building?
- Is the building zoned for the potential uses that might be developed?

**Cost Estimating**
- Predicting costs requires detailed construction documents.
- Each reuse project is different, so it is not possible to compare unit costs across projects accurately.
- Accurately predicting costs during the planning phases of the project is very difficult.

---

with elevators as their only means of vertical transportation are seldom feasible candidates for reuse.

The regular arrangement of a building's structure and services is often unappreciated until changes in the building are needed. Modifications of these two systems are very costly and disruptive to operations. Unfortunately, many industrial companies manage their buildings in a way that results in the irregular arrangement of structure and services.[13] Facility management by incremental adjustment occurs when a firm makes numerous small additions to its buildings to solve immediate problems, which are most often a need for more space. The cumulative effect of this facility management style is usually the same: the building's configuration becomes very irregular and specialized for a limited number of production processes. When a change in product inevitably comes the building cannot feasibly accommodate the change.

Local building codes influence the feasibility of any type of building reuse project.[14] The inflexible requirements of many codes drive up project costs needlessly. Present codes are outmoded and do not address the types of construction represented in older industrial buildings. The prescriptive nature of codes limits innovative solutions to building problems because they tend to protect the status-quo methods of construction. Building inspectors tend to enforce the codes differently for rehabilitation and new construction projects; they tend to be stricter for rehabilitation projects. Code compliance for rehabilitation projects takes longer than for new construction, increasing the time and cost to complete the projects. Typical changes resulting from code compliance are concerned with occupant safety and have little impact on economic feasibility.

Technical knowledge about a vacant building is a valuable asset in determining its reuse feasibility. There is usually little current information available on old buildings. The best, and often only, source is building maintenance persons. If they have had a long association with the building, they will probably be a valuable resource about how the building works, what its problems are, and how it has been maintained.

Because of the high site coverage ratio often associated with older industrial buildings (the proportion of the building's ground floor area divided by the total site area), options for reuse and expansion are limited. The preferred ratio is between 40 and 50%; otherwise there is little land available for expansion, on-site parking areas, and access roads.[15]

Inadequate vehicular access up to and into older industrial buildings is another common site-related problem. Facilities with a high site coverage ratio may not have enough available undeveloped land to allow for the vehicular access needed. This problem can be particularly troublesome for facilities located in central cities, such as those that are being considered for incubator space. Old industrial buildings in central city locations are often built right up to adjacent streets. This limits the area needed to maneuver and park trucks. Incubators usually require multiple access points into the building in order to provide each tenant with a separate loading area.

Regional and community site considerations influence the range of reuse opportunities for industrial buildings. Considerable research has been done on how such regional factors as transportation costs, access to markets, access to labor, and cheap utilities affect site selection.[16] At the community level the issues of zoning, building codes, and surrounding land uses and the relations between the building's previous owner and the community can have a more immediate impact on feasibility.

Cost estimating for reuse projects remains a difficult problem. Predicting costs still requires detailed construction documents because each rehabilitation is different. The idiosyncratic nature of each reuse project makes it difficult to compare costs and to estimate costs during the early phases of the design process.[17]

## IMPORTANT RISK FACTORS

Many of the risks involved in reuse work stem from the inability to predict costs early in the planning phase of the project and to deal with unforeseen problems that arise during construction. There are six important risk factors that should be considered in any reuse project.

1. Roof repair and replacement are the two biggest cost factors in the reuse of single-story industrial buildings.[18] On some projects roof repairs may account for 20% of total project costs as a result of neglect by previous owners and the large roof areas of industrial buildings.

2. Fire safety and occupancy code violations are commonplace in older industrial buildings, especially multistory facilities. Older buildings often lack sprinkler systems, proper fire compartmentation, exit routes, and fire stairs. Upgrading the building to meet more recent codes, "often raises refurbishment costs beyond feasible levels,"[19] and does little to improve the economic performance of a building.

3. Toxic waste and building contamination is a recent addition to the list of important risk factors. Old buildings that housed metal working and plating operations often had large underground and underbuilding storage tanks for the temporary storage of wastes. A careful review of a building's history and a thorough site analysis should identify any evidence of these types of storage.

4. The rehabilitation of an occupied building creates a set of special problems. The occupants must be willing to accept a certain amount of downtime, relocation, and general disruption of their work. It is uncommon to find instances of industrial facility rehabilitation projects with little or no downtime.[20] Construction work done around the occupants is more expensive and an inconvenience for both the construc-

tion workers and the occupants. In general a reuse project should be limited to three phases of construction in order to minimize the number of occupant moves. This would require between 30 and 35% of the building to be vacant to accommodate relocated groups of workers.

5. Construction work done below grade level or on the upper floors of a building is more costly than work done on the grade level of a building. This is particularly true if the work involves structural alterations.

6. Modification of a building's vertical transportation systems is a costly proposition. Typically, older multistory buildings have old freight elevators that are unreliable and slow and have a light load capacity. Addition or replacement of these elevators can cost from $60,000 for a used piece of equipment up to $120,000 to $200,000 for a new elevator.

# EXAMPLES OF INDUSTRIAL REUSE PROJECTS

## Michigan Terminal

Michigan Terminal is a small, capital-intensive manufacturing operation that produces wire terminals for automobiles and appliances. All the company's products are stamped out by large automated presses. Annual sales averaged $5 million in the early 1980s. The firm employs 40 people, most of whom are highly skilled tool and die makers and repairpersons.

In 1978 the company faced a series of operational problems that were related to their facility. The production process flow consisted of 34 different steps; this was considered to be an excessive number and resulted in a high product unit cost. Materials were handled and stored in several different locations including outdoors and at a second site. The quality of the workplace was poor and resulted in the loss of a highly skilled toolmaker. After the loss, the owner vowed that he "would never again lose any employee because of poor quality work environment." The company occupied a total of 20,000 sq ft (1,858.0 sq m) of building area, insufficient given increased sales and production requirements.

The owner began considering whether to build a new facility, relocate to an existing facility, or reuse his existing plant. A decision was made to reuse and expand the existing facility. Tax-free economic development bonds were made available, which saved the owner $250,000 in financing charges over conventional sources. Enough land was available adjacent to the existing site, and the building configuration was amenable to an expansion capable of handling the estimated growth in production for the next 7 to 10 years. Reuse would allow the firm to retain its highly skilled work force and ensure continued high product quality. The reuse project could be done with minimal production downtime so that regular deliveries to customers would not be disrupted.

The reuse and expansion project cost $1.35 million and was completed in 1980. The existing plant composed of four connected buildings was completely renovated. New construction totaling 18,000 sq ft (1,672.2 sq m) added indoor warehouse space for raw materials, scrap, and finished products. An enclosed shipping and receiving area and docks were added, along with a new area for large presses. The entire project took 13 months to complete at a cost of $35/sq ft.

Michigan Terminal received several direct benefits from the project. The total number of steps in the production process was reduced from 34 to 24. The plant manager estimated that this accounted for a significant part of the 5% increase in productivity that occurred in the first year of occupancy. Material handling is now all done inside. Warehousing is consolidated into one area. The quality of the entire facility was significantly improved. Fire insurance premiums were lowered over $8,000/year. From an economic analysis standpoint, the entire project has a payback period of 2.6 years at minimum attractive rate of return adjusted for inflation of 21%.

## Kawneer Building in Niles, Michigan

The second example, a small business incubator in Niles, Michigan, illustrates the strategy of a large building made small. In the summer of 1984, the Kawneer Company closed the last of its facilities in Niles. The city of Niles purchased the Kawneer facilities for $1. Niles is also home to a few large manufacturing firms, some of which had recently been involved in corporate takeovers or sell outs. The community was concerned about further closings or relocations to nearby Indiana. Professionals, labor groups, and members of the city government were worried about the concentration of many jobs with just a few employers. They believed that the city's economic development strategy should focus more on helping create new small businesses that could generate a greater diversity of jobs. The new philosophy would be to "grow your own" instead of trying to recruit big firms to move to Niles.

The Greater Niles Economic Development Foundation (GNEDF) took the lead in working to implement this new philosophy. The first application of the grow-your-own philosophy was on the vacant Kawneer building. The GNEDF foundation commissioned The University of Michigan's Architecture and Planning Research Laboratory to conduct a feasibility study on converting the plant into a small business incubator. The study began with a building and site evaluation. The second and third steps consisted of doing community and market need assessments. The final parts of the project were to develop a marketing strategy as well as a plan and program for operating the incubator.

The Kawneer facility consists of two main buildings: a two-story manufacturing complex totaling 61,000 sq ft (5,666.9 sq m), and a three-story office building with 15,000 sq ft (1,393.5 sq m) of floor space. The manufacturing building is composed of 10 adjacent and connected buildings. The evaluation identified the positive and negative features of both main buildings and made an assessment of how these features affected feasibility.

The evaluation determined that the facility had the basic building characteristics necessary to become a small business incubator. The facility was acquired essentially for free and was in excellent overall condition. The structural system and location of the walls in the manufacturing building were such that it could easily be divided into small parcels of space needed by new businesses. Ceiling clearances averaging 12 ft (3.7 m) were sufficient to accommodate most light manufacturing and warehousing applications. Portions of both buildings were ready for occupancy with only minor modifications and the addition of partitions and drive-in doors.

The community assessment identified whether the needed community resources existed to support an incubator. Resources in the following five categories were assessed: technical assistance and job training; attracting employees; help in relocating business; assistance in new business start-up; and the availability of technical information in local libraries. The assessment indicated that the community had strong resources in the most important area, that of assisting new business start-ups and technical assistance.

The market need assessment examined the existing market conditions, economic constraints on Niles's development, and the profitable and unprofitable market sectors in the local economy. The results indicated that, with few exceptions, firms entering the trade area were small businesses with well under 20 employees. The most actively developing industry sectors included health and allied services, primary metal industries, plastic products, metal stamping, structural clay products, plastic materials and resins, and industrial gases. These findings have been used to help target likely candidates for the incubator.

A six-point marketing strategy was developed to inform new small businesses and people considering starting small businesses of the Niles incubator. Feature stories on key events about the incubator were placed in local and regional papers and on television and radio. A simple brochure and poster were planned for local distribution and prospective tenants. A local business network, which was already active, increased its offerings of free management seminars on how to start a new business. A slide program describing the concepts for the Niles incubator was developed and used in local presentations. A finders' fee program was developed to provide a fee to people who referred tenants to the incubator.

The final component of the project was to develop a plan and program for the start-up and operation of the incubator. The organization governing the incubator is headed by an eight-member community board representing the interests of business, labor, government, and education. The board is responsible for hiring the incubator manager, reviewing tenant applications, and determining the appropriate business support services the incubator will offer. Tenant selection is the most important board responsibility. The potential tenant should have a reasonable likelihood of being profitable within 2 to 3 years. But more importantly, the prospective tenant should not operate a business that could adversely affect existing incubator tenants. For example, a business whose work is dangerous or produces large quantities of toxic wastes should not be allowed in an incubator. New businesses can least afford long periods of downtime caused by the actions of others in a shared facility.

The market research indicated that it would take 3 years to fill the space in the incubator. A three-phase plan was developed to ensure that space was renovated as needed and to minimize the costs of financing the project. Selected portions of the manufacturing and office buildings were to be renovated on the basis of the desirability of space. Special rental incentives were to be offered to firms considering the less desirable space.

The economic analysis of the project indicated that it would be profitable in its second year of operation if the rental rate was between $2.75 and $3.00/sq ft. Financing was from surplus city funds and was to be provided both as a grant and a loan. Total projects costs were estimated to be $350,000 spread over 3 years. In return for the city's contributions, the incubator was expected to generate 120 new jobs by 1988 with a $2.5 million payroll. The incubator began operating in the spring of 1985 with three new businesses as tenants. As Table 15-2 indicates, the city and community are much better off converting the Kawneer plant into an incubator than they would have been selling the facility outright.

## NOTES

1. "Five Case Studies of Displaced Workers," *Monthly Labor Review* (June 1964), pp. 663–670; Stephen S. Mick, "Social and Personal Costs of Plant Shutdowns," *Industrial Relations* 14 (May 1975), pp. 203–208; Don Stillman, "The Devastating Impact of Plant Relocations," *Working Papers* (July 1978), pp. 42–52; Barry Bluestone and Bennett Harrison, *Capital and Communities: The Causes and Consequences of Private Disinvestment,* (Washington, D.C.: The Progressive Alliance, 1980); Ed. Kelly and Lee Webb, eds., *Plant Closings: Resources for Public Officials, Trade Unionists and Community Leaders,* Washington Conference on Alternative State and Local Policies, 1979; Iver Peterson, "Firestone Tire Plans to Close 6 Plants Idling 7,625," *The New York Times* (20 March 1980), p. 1.

TABLE 15-2
**Economic benefits of selling the Kawneer Plant versus converting it into a small business incubator**

| | PRESENT WORTH | YEAR 1 | YEAR 2 | YEAR 3 | YEAR 4 | YEAR 5 |
|---|---|---|---|---|---|---|
| City sells Kawneer for $465,000 and invests this at 11% | $465,000 | 516,150 | 572,926 | 635,948 | 705,902 | 783,552 |
| Kawneer plant converted into an incubator [a] | $697,741 | 664,200 | 933,336 | 1,008,003 | 1,088,643 | 1,175,735 |
| Value of jobs generated by firms in the incubator [b] | . | 700,000 | 1,587,600 | 2,449,440 | 2,645,280 | 2,856,960 |

[a] Assume the value of the building increases at 8 percent annually because it will be occupied. Plus the total investments of $350,000 in years one and two will increase the value.
[b] The number of jobs are 40 in year one, 84 in year two, and 120 in the out years. Average salary starts at $17,500 in year one and increases 8 percent annually.

2. "The Effects of Rehabilitation on the Corporate Balance Sheet," *Hoffman Report* 2, no. 1.

3. Lincoln C. Jewett, "How to Market Surplus Real Estate," *Harvard Business Review* (January 1977), pp. 7–8.

4. David S. Levy, *Rehabilitation of Obsolescent Industrial Facilities for Continued Industrial Uses* (Ann Arbor, Mich.: University Microfilms, 1982), pp. 75–82.

5. Levy, *Rehabilitation of Obsolescent Facilities*, 75–78; Harrison and Bluestone, *Capital and Communities*, p. 228; Sam Allis, "States Pay Dearly, Gain Little in Competition to Lure Industry," *The Wall Street Journal* (1 July 1980), p. 23; Roger J. Vaughan, "How Effective Are State Tax Incentives?" *Commentary* 4 (January 1980), p. 5.

6. Vaughan, "How Effective Are State Tax Incentives?" p. 5.

7. Allis, "States Pay Dearly," p. 23.

8. Gene F. Schwind, "Appliance Park-East: Total Design for Material Handling," *Material Handling Engineering* 26 (September 1971), pp. 85–87; Thomas M. Harrison, "A Breakthrough in Job-Shop Layout," *Modern Materials Handling* 26 (July 1971), pp. 30–33; "Superb Material Control Despite Layout Handicaps, " *Modern Materials Handling* 34 (July 1980), pp. 62–69; George A. Weimer, "How to Get the Most Out of Your Plant Through Good Design," *Iron Age* 11 (March 1981), pp. 48–53.

9. P.A. Stone, *Building Design Evaluation: Costs-in-Use,* (London: E. and F. Spon, 1967), p. 16.
10. Hildebrand Frey, "Building Conversion: A System for the Prediction of Capital Cost," in *Building Conversion and Rehabilitation: Designing for Change in Building Use,* ed. Thomas A. Markus (London: Newnes-Butterworths, 1979), p. 31.
11. Levy, *Rehabilitation of Obsolescent Industrial Facilities,* p. 124.
12. Frank Eul, "Industrial Refurbishment: 2 Feasibility," *Architects Journal* 8 (August 1982), p. 50.
13. Levy, *Rehabilitation of Obsolescent Industrial Facilities,* p. 170.
14. James G. Gross, James H. Pillert, and Patrick W. Cooke, *Impact of Building Regulations on Rehabilitation-Status and Technical Needs,* U.S. Department of Commerce, National Bureau of Standards Technical Note 998 (Washington, D.C.: U.S. Government Printing Office, 1979), pp. 1–5; John Habraken and Ann Beha, *An Investigation of Regulatory Barriers to the Re-Use of Existing Buildings* (National Bureau of Standards, 1978), pp. 6, 16, 34.
15. Eul, "Feasibility," p. 49.
16. F.E. Ian Hamilton, ed., *Spatial Perspectives on Industrial Organization and Decision-Making,* (New York, John Wiley and Sons, 1974); Barry M. Moriarity, *Industrial Location and Community Development* (Chapel Hill, N.C.: The University of North Carolina Press, 1980); Roger Schmenner, "Choosing New Industrial Capacity: On-site Expansion, Branching, and Relocation," *Quarterly Journal of Economics* (August 1980), pp. 100–104.
17. Frey, "Building Conversion," p. 31; Public Technology Inc., "Recycling of Obsolete Buildings," (Washington, D.C.: Urban Consortium, 1977), p. 6.
18. Eul, "Feasibility," p. 50.
19. Eul, "Feasibility," p. 50.
20. Levy, *Rehabilitation of Obsolescent Industrial Facilities,"* pp. 200–209, 249–261.

# 16

## CHAPTER SIXTEEN

# ALEPH PARK AND APPLIED COMPUTING DEVICES: CASE STUDIES IN HIGH-TECH FACILITIES DESIGN

## *HARRY A. EGGINK AND ROBERT J. KOESTER*

The following synopsis is extracted from a larger body of publications. The work resulted from a university/government/industry research partnership that centered on high-tech facility design for a case study company that was to be located in Aleph Park, a prototype industrial setting. This 106-a tract of land in Terre Haute, Indiana, is presently being developed according to the master plan and design guidelines originating from the research program summarized in this chapter. Construction of the building to house the case study company—Applied Computing Devices—is complete and in use. Other tenants for the park are being sought.

Included in the following overview are a brief historical sketch of industrial development in the United States, a discussion of the uniqueness of contemporary high-tech facility design, a review of the research methodology, and a summary of the research findings and implementational recommendations for the park. A final facility design commentary is provided at the close of the chapter.

## HISTORICAL SURVEY

The history of industrial/manufacturing facilities development and design has evolved along several convergent paths. These paths reflect physiographic, technological, and sociological influences. A chronological sketch of this evolution would include both siting and building design examples. Siting factors are evident in the location of mills near water, the adjacency of factories to canal or rail shipping, and more recently, the placement of industrial parks near the nation's interstates and airports. Today, with the potential for satellite communication linkage and on-site energy autonomy, site selection criteria and site development as concomitants to facility design are about to take the next evolutionary step. In fact, the unique function of high-tech industry further complements this change. With the growth of the high-tech field, industry is moving away from its traditional function as a principally manufacturing-centered activity. The dictates of assembly line process, the need to tap raw power, and the availability of transportation resources are giving way to new facility design criteria. Emphasis falls on a company's research and development processes in which creativity and human interaction are largely "the product." The research park ambience—a setting in which communication is critical and natural resources are valued more for their impact on both worker and company productivity than for their advantage as a transportation link or mechanical power source—is the model. In this context, sites can be selected for their physical attractiveness as work settings and their usefulness in meeting more modest but very sophisticated energy needs.

## Familiar Images

If we look at the chronology of industrial development in more detail, we can understand the importance of this shift and categorize the prime factors of each step in the evolutionary trend.

### EARLY AMERICAN MILL
The early American mill was one of the first production-specific constructs in American architecture. The mill was sited for best use of available power, and a water's fall was frequently augmented with dam or sluice construction. The motive power of these mills was an effective, well-suited source of energy for the grinding of grain. Appearance of the mills marked the end of an almost totally agrarian society and the beginning of the impact of the Industrial Revolution (Figure 16-1a).

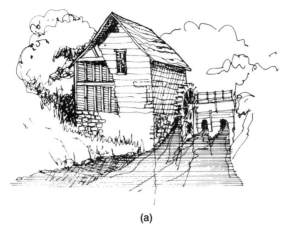

(a)

**Figure 16-1.** Historical development of industrial/manufacturing facilities. (a) Early American mill. (b) Post–Civil War industrial facility (mid 19th century). (c) Mass production factory (early 20th century). (d) Industrial park (mid 20th century). (e) The next step.

## POST–CIVIL WAR FACTORY

By the mid-nineteenth century, the introduction of steam power completely changed the complexion of the industrial facility. The linear axis of central power shafts for the belt-driven machinery dominated building geometry. This same power source made multistory construction useful. Steam power also fostered construction of an extensive rail network, which in turn redefined site dependency. Access to rail corridors, in effect, urbanized the factory (Figure 16-1b).

## MASS PRODUCTION

At the beginning of the twentieth century, widespread use of mass production techniques again affected major changes in factory form, location,

(b)

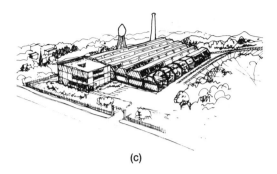

(c)

and functional emphasis. Horizontal layout of the assembly line and the use of fork lifts for movement of goods dictated sprawling, high bay construction. Access to rail and trucking networks became an additional important influence in terms of either distribution of goods or access to raw materials (Figure 16-1c).

## INDUSTRIAL PARK
The mid-twentieth century was a boom time for a new phenomenon—the industrial park—a collection of factory buildings in the middle landscape, the suburbs. The facilities located in these settings were designed primarily for the flexibility and growth of the industry with some modest considerations given to quality of the work environment. The industrial parks' locations outside the cities were dictated in part by a desire for ready access to the new interstate system (Figure 16-1d).

## THE NEXT STEP
In light of this history, which shows such strong influence of energy, transportation, and communication in the linkages of activity, building, and context, the natural question is: What is the likely confluence of present trends that will give form to new industrial facility design? How will our design responses to contemporary technological, physiographic, and/or sociological factors contribute to the next evolutionary step in industrial facility design? (Figure 16-1e).

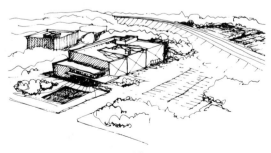

(d)

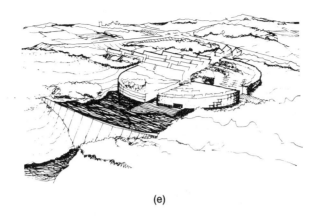

(e)

Many factors are at play. And while it would be impossible to completely anticipate or accurately predict developments that will occur, several immediate trends have clear implications. "Demassification" of industry with increased emphasis on the quality of work environment will redefine the meaning of "industrial setting."

Contemporary industrial development will focus evermore on the link of worker satisfaction with corporate productivity. As in the case of the early mill, building and site will be highly integral; site developments, building form, and material selection, however, will be manipulated for their contribution to enhanced worker environment—the sense of "place" as a setting for human interaction and communication.

Demassification will also drastically modify the role of energy considerations. Raw power will fall away as the issue. The impact of energy will be measured less in terms of total consumption and more in terms of its quality and purpose as an end use—i.e., affecting the character of the luminous and thermal environment. A resulting spin-off of the shift in magnitude of energy demand will correspondingly affect the need for, and character of, facility centralization. As in the early mill example, contemporary industry can become more energy autonomous—disconnected from the grid and more self-reliant—perhaps even more "rural" in its setting.

## DEFINING HIGH-TECH

Demassification, as a concept, is principally tied to high-tech industry. As a catch phrase, "high-tech" is used rather generally to describe or refer to all "computer-based" or "computer-related" industry. Many people in the field, in fact, actually dislike the term, for its generalization and lack of definition. The following section further defines high-tech and, in particular, extends that definition as applied to the case study company, which served as catalyst for the research program discussed.

## The National Overview

A recent article recounting the findings of a 1-day Urban Land Institute conference attempts to define criteria for siting and accommodation of the high-tech firm: "Employee lifestyle needs, access to major universities, and little government interference weigh as heavily in the selection of a facility [read operational setting] for a high tech company as traditional corporate location concerns."[1]

The purpose of the conference, which was sponsored by William G. Rouse, III, was to discover "just how different these new buildings will be in terms of location, site configuration and infrastructure, amenities, and actual building systems and design."[2] The identification of trends that emerged from the meeting included the following:

- Competition for key technical personnel (which) makes concern for the overall quality of life central
- University access—to meet the needs of both the company and the employee
- Management's desire for a business climate free of excessive regulation and unnecessary costs
- Linkages and clusters, the ability of these firms to settle near one another
- Building systems that will move closer to capacity for integrated provisions of power; heating, ventilating, and air-conditioning; security; and telecommunications
- Flexibility for future changes

Each of these deserves examination, but one in particular takes on special importance—that is, the necessity for *clustering*.

*Firms now cluster because they need to be near specialized vendors and suppliers. They need to be near a larger employment pool that is provided by a number of peers and competitors in a particular area.*

*Importantly, clustering increases the aggregate "weight" of these firms, which improves their ability to influence state and local jurisdictions in zoning, tax, and regulatory affairs.[3]*

The "definition" of high-tech arising from the comments above hinges on concerns that can be said to be "regional," if not "global," and are cast from a management point of view. The thrust of discussion is built from an economic development, "how can we attract and retain" inventory of community resources perspective?

In a similar profiling of the "high-tech economic development opportunity," the U.S. Congress Joint Economic Committee issued a staff report that used a survey method to determine prioritization of factors affecting location choices—both regionally and within regions (Tables 16-1, 16-2). This study, as well, is rather administratively based.[4] It emphasizes those broader concerns that would influence or affect governance, taxation, and worker availability—those aspects of a community's resources that might serve as arms of persuasion for retention or relocation. But in neither case do these studies survey or address the workers themselves. When such a survey was made, additional important qualifiers were revealed.

## The Case Study Company

A worker survey of the case study high-tech company Applied Computing Devices (ACD) was conducted. The intent was to better understand the reason for, and effect of, its location in Terre Haute, Indiana, and to measure the programming criteria critical to the proposed new facility design. As noted in volume 6 of the Aleph Park report: "All six of the factors which were job and employer-related were rated among the seven highest in importance. [Table 16-3.] There was no significant difference between the ratings of professional and technical-skilled employees."[5]

TABLE 16-1
**Factors that influence the regional location choices of high-technology companies**

| RANK | ATTRIBUTE | % SIGNIFICANT OR VERY SIGNIFICANT[a] |
|------|-----------|-------------------------------------|
| 1 | Labor skills/available | 89.3 |
| 2 | Labor costs | 72.2 |
| 3 | Tax climates within the region | 67.2 |
| 4 | Academic institutions | 58.7 |
| 5 | Cost of living | 58.5 |
| 6 | Transportation | 58.4 |
| 7 | Access to markets | 58.1 |
| 8 | Reg'l regulatory practice | 49.0 |
| 9 | Energy costs/available | 41.4 |
| 10 | Cultural amenities | 36.8 |
| 11 | Climate | 35.8 |
| 12 | Access to raw materials | 27.6 |

[a] Respondents were asked to rate each attribute as "very significant, significant, somewhat significant, or not significant" with respect to their location choices. The percent of very significant and significant responses were added together to obtain an index of overall importance.

**TABLE 16-2**
**Factors that influence the location choices of high-technology companies within regions**

| RANK | ATTRIBUTE | % SIGNIFICANT OR VERY SIGNIFICANT |
|------|-----------|-----------------------------------|
| 1 | Availability of workers | 96.1 |
| | Skilled | 88.1 |
| | Unskilled | 52.4 |
| | Technical | 96.1 |
| | Professional | 87.3 |
| 2 | State and/or local government structure | 85.5 |
| 3 | Community attitudes toward business | 81.9 |
| 4 | Cost of property & construction | 78.8 |
| 5 | Goods transportation for people | 76.1 |
| 6 | Ample area for expansion | 75.4 |
| 7 | Proximity to good schools | 70.8 |
| 8 | Proximity to recreational and cultural opportunities | 61.1 |
| 9 | Good transportation facilities for materials and products | 56.9 |
| 10 | Proximity to customers | 46.8 |
| 11 | Availability of energy supplies | 45.6 |
| 12 | Proximity to raw materials component supplies | 35.7 |
| 13 | Water supply | 35.3 |
| 14 | Adequate waste treatment facilities | 26.4 |

The indications from this "worker profiling" are many:

• The job is the drawing card

• Career advancement was rated important by both older and younger workers

• Work atmosphere was slightly more important for technical-skilled than for professional employees

• Those who rated professional challenge high, interestingly, also rated salary high

• Salary was not the highest of the job-related factors

• Work environment was rated high by two thirds of the respondents

A generalization that can be made from comparing the three studies cited is that there does seem to be some distinction between *an administrative view* that firms will locate in order to "draw employees whose interests center on (regional) environmental amenities" and an *employee view* that firms ultimately "draw on those whose interests are professional." In other words, *"Firms that make themselves attractive* to employees, professional or technical, are actually following a prescription for success."[6]

**TABLE 16-3**
**Factors in the decision of employees to locate in Terre Haute and West Central Indiana**

| RANK | ATTRIBUTE | % CRUCIAL OR IMPORTANT |
|---|---|---|
| 1 | Job opportunity with this employer | 88 |
| 2 | Proximity to family | 78 |
| 3 | Career advancement opportunity | 75 |
| 4 | Pleasant work atmosphere | 75 |
| 5 | Challenging professional responsibilities | 70 |
| 6 | Salary | 69 |
| 7 | Physical work environment | 65 |
| 8 | Overall cost of living | 63 |
| 9 | Longtime resident of area | 61 |
| 10 | Proximity to continuing education opportunties | 56 |
| 11 | Housing (owner): price | 35 |
| 12 | Proximity to recreational opportunities | 32 |
| 13 | Housing (owner): price | 32 |
| 14 | Quality of education public schools | 31 |
| 15 | Geographic location | 26 |
| 16 | Rental housing: selection | 25 |
| 17 | Rental housing: price | 23 |
| 18 | Entertainment opportunities | 23 |
| 19 | Local taxes | 19 |
| 20 | State taxes | 19 |
| 21 | Proximity to a large city | 19 |
| 22 | Job opportunities available | 17 |
| 23 | Availability of private schools | 3 |

As mentioned, this case study survey was used also to extract a meaningful understanding of the specific ACD work environment programmed for inclusion in Aleph Park (Table 16-4). The extent to which ACD represented a break with industrial facility design history, the fact that the company proved to be an example of general trends and the fact that Aleph Park provided a prototyping opportunity laid the foundation for the research discussed herein.

## NEED FOR THE STUDY

Programmatically, ACD fits the high-tech categorizations listed earlier. Broadly, it is in Alvin Toffler's words a "demassified industry."[7] The program for ACD reveals the large percentage (nearly two thirds) of research and development, management, and marketing personnel/space allocation; this is in contrast to the traditional "massified" or principally manufacturing-based industry as found in the opening historical sketch.

Aleph Park was initially conceived as having 10 industries in location. The 106-a tract has several natural amenities, is located near the interstate

**TABLE 16-4**
**Program of spaces for the new ACD facility**

|  | POP. CT. | PROG. AREAS |
|---|---|---|
| *General and Administrative (G&A)* | | |
| President | (1) | 550 |
| Controller | (1) | 500 |
| CPA | (1) | 500 |
| Clerical | (4) | 750 |
|  | | 2250 ft |
| *Engineering* | | |
| Engineering | (3) | 2000 |
| Test Engineering | (3) | 500 |
| Research and development | (25-30) | 6000 |
|  | | 8500 ft |
| *Manufacturing* | | |
| Assembly I | (4) | 3000 |
| Assembly II | (4) | 3000 |
| Repair/test | (5-7) | 600 |
| Final test | (3) | 3000 |
| Product inventory control | | 200 |
| Quality assurance | | 500 |
| Purchasing | | 500 |
|  | | 10,800 ft |
| *Marketing* | | |
| Marketing services | | 1700 |
| Sales | | 1500 |
| Product support | (6) | 2000 |
| Product manager(s) | (4) | 1000 |
| Project manager(s) | (2) | 500 |
|  | | 6700 ft |
| *Miscellaneous* | | |
| Break areas, restrooms | | |
| Mech. | | 2000 |
|  | | 2000 ft |
| Grand Total | (80 +/−) | 28,250 ft |

and airport, and is on the periphery of an existing city infrastructure. From site, to building, to workstation, the precedents were few and the opportunities extensive. Having a real project amplified these opportunities and reinforced the need for the research. The uniqueness of the government/industry/university partnership in this project provided the broadest possible base for the research work. To that end, final reports on this study included extensive documentation of the research methodology, the design explorations, and the final recommendations. The recommendations included a master plan, design guidelines, and design management policies and procedures. While the final recommendations for development of Aleph Park continue to be implemented, it is important to note that the scope of this study and the final deliverable were not structured to replace a professional design service; rather, the intent of the study was to circumscribe the range of concerns and the complexity of issues involved in such a project—from the scale of site to the scale of individual workstation.

## THE RESEARCH METHODOLOGY

This study required a careful orchestration of analytical and synthetic work. On the one hand, there was a need to derive an understanding of influences, forces, or constraints that existed as a result of the realities of the case study—the client, the program, the site. On the other hand, there was a need to outline more universal parameters that could apply in somewhat different client/site region contexts. To meet this obligation of addressing the "particular" and the "universal," a hierarchy of information scales were set. At the lowest level, information was gathered or presented in *issue* form. In addition, problem *models* were used to assess, integrate, or otherwise show the influence of the issues of concern. Finally, design *strategies* were presented that covered a range of alternative approaches to the energy and communication needs of this industry type. The study was also organized in terms of three different scales of focus: *site, building,* and *space.* These informational levels and scales of focus offered the needed brackets of control for the analytical and design work.

### Issues

As a first-order listing of concerns, this study documented pertinent issues. These were the rather conventional points of reference used in the normal design process of architects, landscape architects, and planners. The issues were both universal in the form of "an inventory" and problem specific in the form of "an assessment."

For example, at the site scale the issues of topography, climate, vehicular access, and solar access were inventoried (Figure 16-2). At the building scale, thermal loads, daylighting, structures, mechanical services, and construction materials were examined (Figure 16-3). And finally, at the spatial scale, human interaction, equipment, and square footages were accounted for (Figure 16-4).

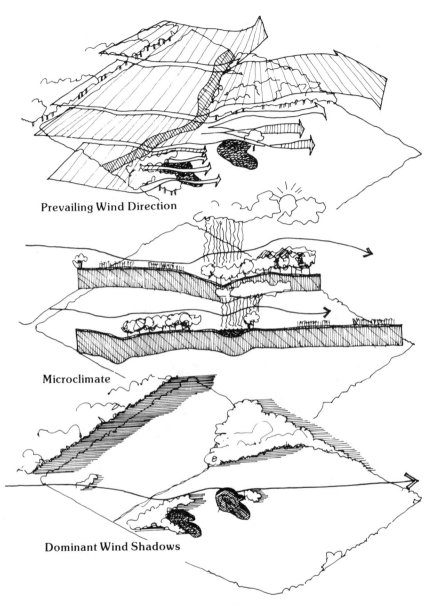

Prevailing Wind Direction

Microclimate

Dominant Wind Shadows

**Figure 16-2.** *Issues:* site scale—topography, climate, vehicular access, solar access. The property contains five microclimate zones, influenced primarily by solar exposure, wind exposure, and on-site humidification factors (vegetation and water).

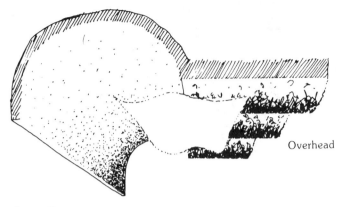

Overhead

**Space Formed**

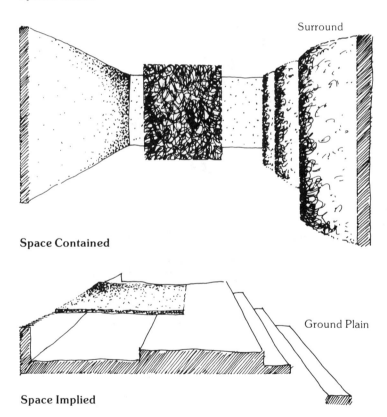

Surround

**Space Contained**

Ground Plain

**Space Implied**

**Figure 16-3.** *Issues:* building scale—thermal loads, daylighting, structures, mechanical services. Daylighting, energy, structures, and mechanical services in building design are discussed in terms of two frameworks: building component zones and spatial patternings of people.

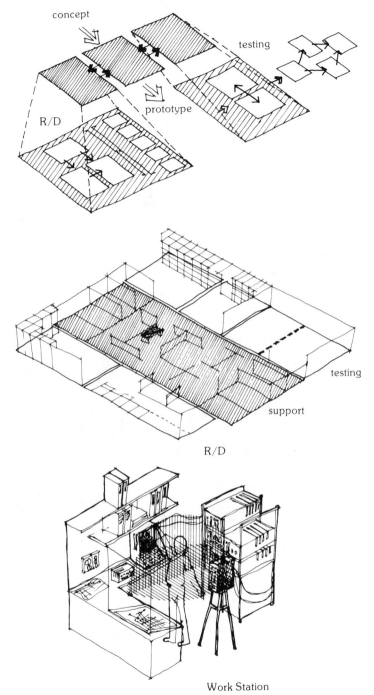

concept

testing

prototype

R/D

testing

support

R/D

Work Station

**Figure 16-4.** *Issues:* spatial scale—human interaction, equipment, square footages. The framework used for program development was broken into three parallel streams of inventory/analysis/documentation: it included work habits and worker association, thermal luminous and acoustic criteria, and appliances and furnishings associated with each task.

# Models

While conducting such informational inventory can be a lengthy and sometimes arduous process, the "payoff" comes through the assessment of implications, constraints or potential embodied in the interrelationships of these factors for the specific project (Figure 16-5).

In order to gain some control over the interrelationships, it was helpful to "build up" the understanding through selected studies—that is, the making of models of interrelationships. For the sake of this study such models were presented by scale.

Direct Relationship

Indirect Relationship

Two primary options for vehicular/building/site relationships are clear: the direct relationship and the indirect relationship. Several generic interpretations are diagramed below.

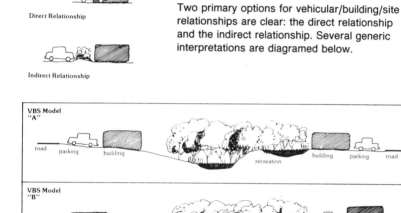

VBS Model "A"

road   parking   building   recreation   building   parking   road

VBS Model "B"

road   building   parking   recreation   parking   building   road

VBS Model "C"

road   parking building   recreation   parking building   road

VBS Model "D"

road   building parking   recreation   building parking   road

**Figure 16-5.** Interrelationships of variables for specific project.

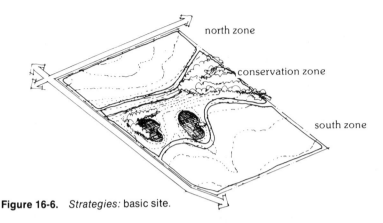

**Figure 16-6.** *Strategies:* basic site.

## Strategies

As a final step in preparing this work, a series of three strategies were out-lined. These were broad-based sketches of organizational themes. Each in turn was explored with the intent of establishing its validity: broadly, the three strategies studied were the "traditional," the "infrastructure," and the "energy park."

### TRADITIONAL
Under this strategy, the site was assumed to be subdivided into building plots. Each plot was to be separately developed as a single unit linked to the industrial park by a unified landscape development plan, with orientation guidelines. The size of the plot was to depend on the space and future growth needs of the individual industry (Figure 16-7).

### INFRASTRUCTURE
Under the infrastructure strategy, the northern and southern halves of the park were to be treated as independent but unified parcels accommo-dating up to four or five industrial facilities. These were to be "linked" or organized around a fabric, matrix, or corridor of common infrastructure. The common fabric would include shared computer and communication systems, utility corridors, energy components, physical facilities, and out-door and landscape facilities (Figure 16-8).

### ENERGY PARK
Under the energy park strategy, the "whole" site was to be considered as a single unit of development. The strategy was to create a total energy net-work shared by all the corporate facilities, which would provide the ability to proceed toward self-sufficiency. The park would utilize passive energy concepts as well as more active energy conversion technology in order to produce energy-efficient environments (Figure 16-9).

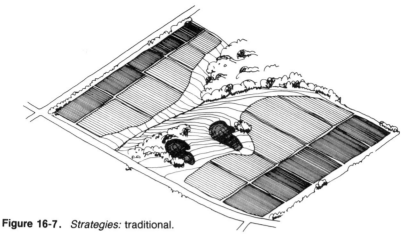

**Figure 16-7.** *Strategies:* traditional.

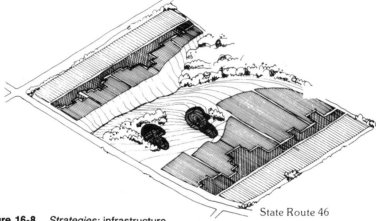

**Figure 16-8.** *Strategies:* infrastructure.

State Route 46

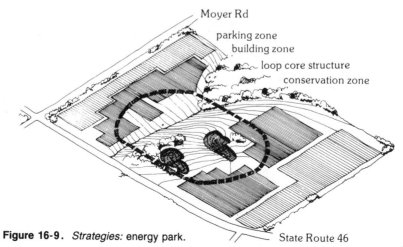

Moyer Rd

parking zone
building zone
loop core structure
conservation zone

**Figure 16-9.** *Strategies:* energy park.

State Route 46

**329**

## SAMPLE EXPLORATIONS

The design studio served as the principal vehicle for exploration of the architectural potential of site, building, and workstation integration. The variety of responses evidenced in the student work highlighted the richness of building/site integration that can be achieved within the constraints of the facility program and site development strategies (see Figure 16.10). As a result of many student studies and the interaction with the case study clients, the final master plan, design guidelines, and design review policies and procedures were established.

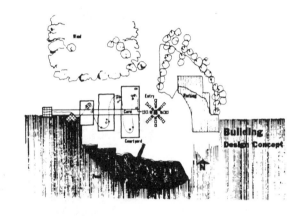

(a)

**Figure 16-10.** **(a to c)** Sample design explorations.

(b)

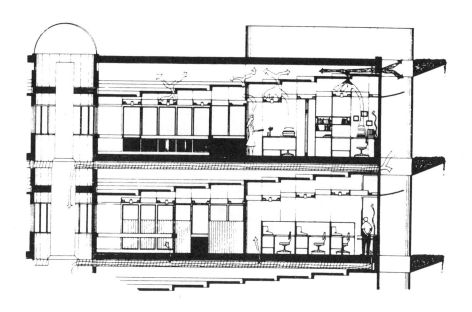

(c)

331

## FINAL RECOMMENDATIONS

### The Master Plan

The plan shown in Figure 16-11 was the final proposal for the development of Aleph Park. The structure of this scheme was based on maximizing the site amenities and providing a unified final form to the site as a whole. In addition, the attempt has been to simplify the means by which definition of tenant/community "ownership" can be made. Traditional site plot boundaries were shown as superimposed on, but integral to, specific common site features and amenities. Simplified zoning of public/semipublic/semiprivate/private boundaries were also used to pattern the development areas.

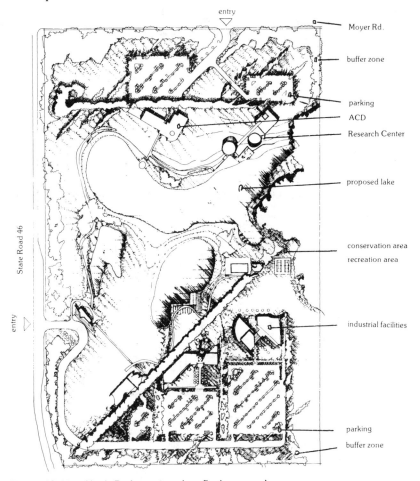

**Figure 16-11.** Aleph Park master plan: final proposal.

# The Design Guidelines

Design guidelines were developed principally for use by the Aleph Park Corporation. The interest was to structure a means for achieving design consistency among the different tenant facilities. These guidelines exceeded those prescriptions or designations normally associated with site plots and actually addressed issues of physical design. Included in the guidelines were minimum requirements or outright restrictions on buildings, paving, vegetation, lighting, signage, and site furniture. Energy performance was also a featured subject of the detailed design criteria (Figure 16-12).

Those criteria specifically addressing building design were written to be more open to or accommodating of the vagaries of program, client, designer; those criteria associated with the landscape tended to be more specific and restrictive. The goal of this technique was to "use" the natural features of the landscape to tie together and provide a consistency to the park as a whole.

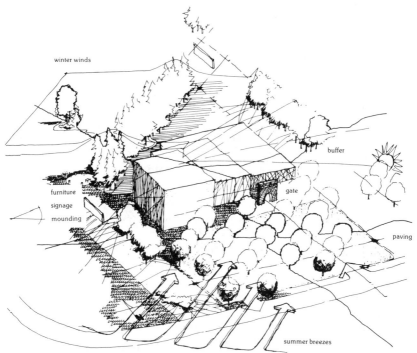

**Figure 16-12.** Design guidelines.

## The Design Policies and Procedures

Design review policies and procedures have been provided as a means to assure orderly interaction between Aleph Park Corporation and prospective tenants of the park. As a policy, all development proposals for Aleph Park must be reviewed for compliance with the design criteria set forth in the design guidelines. It is intended that the Board of Directors of Aleph Park Corporation serve as the body for such review.

The procedures by which such review is to occur are set out as a sequence of events that parallel the normal design process for new building/site development work. There is to be a predesign conference in which "mutual design objectives, the characteristics of the particular parcel and technical issues related to design review procedure" are to be discussed.

Following initial work by the architects, landscape architects, and other consultants chosen by the prospective park tenant, a schematic design review is to be held. Sufficient information then is to be provided to enable the board to substantially evaluate the proposed facility and provide specific comment and recommendations within a short time. Preliminary plan approval is required next. At this stage, very detailed information is submitted, and again the board is to provide comments and recommendations. As the last step in the review process, the board must give final plan approval based on contract documents as submitted by the prospective tenant. The intent of this review is to assure compliance of the final design with that previously approved.

## COMMENTARY

The specific recommendations to come from this research emphasize the range of concerns and the complexity of issues involved in such a project—from the scale of site to the scale of individual workstation. The master plan, design guidelines, and design management policies provided the instruments needed to affect a park that capitalizes on available site amenities to meet the high-tech facility design/worker setting criteria.

The student work suggested the breadth of opportunities in building form, organization, and detailing that is possible for the new industry type. Moreoever, this building/site response can be somewhat independent of the many workstation needs and their constantly shifting arrangement.

The documents referenced in the opening paragraphs record the translation of these concerns into a multiscaled and multifactored process of study of "energy" and "communication" as the problematic themes of "high-tech" facility design. The principles derived and the examples shown were intended to assist the Indiana Department of Commerce in working with the city of Terre Haute and other such communities throughout the state in developing strategies for the recruitment and accommodation of computer-based, high-tech industry based on the critical need of worker satisfaction with the immediate work environment.

At the minimum, the study opens a door to the realization that successful high-tech facilities must adhere to the more timeless architectural principles in seeking to obtain their structure. Energy and communication are two themes for this realization. Moreover, the high-tech industrial shift can contribute to a clear realignment of architectural principles. And while the findings of this research did not imply a rediscovered architectural style, it did suggest that "criticality" of the work environment can be the basis of an insightful and inventive (as opposed to fashioned) architecture.

## ACKNOWLEDGMENTS

Through the coordinative efforts of Dean Robert Fisher, the authors were able to work with nearly a dozen faculty and over 40 students in the architecture and landscape architecture programs at Ball State University. Particular thanks are due to Michel Mounayar for his extensive input, Stan Mendelsohn for his timely support, and the many faculty whose expertise was critical to the work: Omar Faruque, Robert P. Meden, J. Paul Mitchell, J. Rodney Underwood; we would also like to thank Bruce Kieffer and Dave Schoen for their contribution of the CERES-2 (now CERENET®) computer simulations. Finally, our biggest thanks go to our students who explored the design strategies and uncovered the many architectural options, which reinforce the fundamental role of communication and energy as principal design themes of the high-tech facility.

## NOTES

1. Ronald Derwen, "Why High Tech Firms Gravitate Toward Site Clusters," *Facilities Design and Management* (July/August 1983), pp. 60–63.
2. Ronald Derwen, *op cit.*
3. Ronald Derwen, *op. cit.*
4. U.S. Congress, Joint Economic Committee, *Location of High Technology Firms and Regional Development.* A staff study prepared for the use of the subcommittee on Monetary and Fiscal Policy (June, 1982), p. 23.
5. *Aleph Park: A Case Study in Design: Computer Based Hi-Tech Industrial Site Planning,* (Muncie, Ind.: Ball State University, 1984). Also of inter-

est: vol. 1, *A Design Resource Book;* vol. 2, *The Case Study Program;* vol. 3, *Linking Education and Research;* vol. 4, *Master Plan: Aleph Park;* vol. 5, *Design Guidelines: Aleph Park;* vol. 6, *Analysis and Assessment Primer.*

**6.** Ibid.

**7.** Alvin Toffler, *Future Shock* (New York: Bantam, 1980).

# ACRONYMS AND ABBREVIATIONS

**AI:** artificial intelligence
**AO:** automated office
**AR:** artificial reality
**ANSI:** American National Standards Institute
**ASHRAE:** American Society of Heating, Refrigeration and Air Conditioning Engineers
**AS/RS:** automatic storage and retrieval systems
**ASTM:** American Society for Testing and Materials
**AT&T:** American Telephone and Telegraph
**BAS:** building automation systems
**BOCA:** Building Officials and Code Administrators
**BOSTI:** Buffalo Organization for Social and Technological Innovation
**CAD/CAM:** computer-aided design/computer-aided manufacturing
**cfm:** cubic foot per minute
**CIS:** Center for Integrated Systems (Stanford University)
**CRT:** cathode ray tube
**db:** decibel
**DDC:** direct digital control
**DNA:** deoxyribonucleic acid
**FAX:** facsimile transmission
**fc:** footcandle
**FTS:** Federal Telecommunications System (U.S.A.)
**GSA:** General Services Administration
**HVAC:** heating, ventilating and air-conditioning
**IB:** intelligent building
**IBM:** International Business Machines
**ISDN:** integrated services digital networks
**ISO:** International Standards Organization
**IT:** information technology

**LAN:** local area network
**LED:** light-emitting diode
**MCC:** Microelectronics and Computer Technology Corporation
**MCNC:** Microelectronic Center of North Carolina
**MIS:** management information system
**MIT:** Massachusetts Institute of Technology
**NASA:** National Aeronautics and Space Administration (U.S.A.)
**NSF:** National Science Foundation
**OA:** office automation
**ORBIT:** Organizations, Buildings and Information Technology
**PARC:** Palo Alto Research Center (Xerox Corporation)
**PAQ:** Position Analysis Questionnaire
**PBS:** Public Buildings Service (of the GSA)
**PBX (PABX):** private (automated) branch exchange
**PC:** personal computer
**PI:** principal investigator (of a research project)
**POE:** post-occupancy evaluation
**psi:** pounds per square inch
**RAM:** random access memory
**ROM:** read only memory
**SUNY:** State University of New York
**TLV:** threshold limit value
**UPS:** uninterrupted power supply
**VAV:** variable air volume
**VDT:** video display terminal; visual display tube; video data terminal

# GLOSSARY

**American Society for Testing and Materials (ASTM):** Now the largest standards-setting organization in North America, with international membership.

**anthropometrics:** Science concerned with the description and measurement of the dimensions of the human body.

**appropriation:** 1) In a budget, the allocation of specific funding to a project. 2) In the context of technological change, the process enabling the intended users of new technologies to get to learn them and make them their own.

**architecture:** 1) The art or science of building. 2) In computing, refers to the logical structure and concept underlying a design (e.g., software architecture).

**artificial intelligence (AI):** A recent field of science at the overlap between computer science and cognitive psychology dealing with the digital modelling and mimicking of human "intelligent" functions. May call upon artificial sensory perception (e.g., in robotics) and expert systems.

**artificial reality (AR):** The production of highly realistic, but totally artificial, computer-generated images, using advanced image-processing technologies.

**automated office (AO):** Shorthand for an office organization where a substantial proportion of tasks have been automated and make use of new technologies. Sometimes also called: "office of the future."

**automatic storage and retrieval systems (AS/RS):** Used in manufacturing plants to put away and to fetch from storage automatically goods and parts while updating the inventory database.

**base isolation:** Used in the design of foundations of buildings in seismic regions. Located between a column and its footing, it prevents seismic forces from reaching and damaging the superstructure.

**bending radius:** For main cables, the minimum radius allowable to change direction in the cable layout.

**biomechanics:** Science concerned with the description and measurement of the human body in motion, including strength, range of motion, reach envelopes, etc.

**bug:** A malfunction in the intended and expected functioning of a computer program. Removing bugs, or debugging, is an important and often time-consuming activity for programmers.

**building automation systems (BAS):** Computer systems that enable the automatic control of various building systems by direct digital control (DCC).

**building to land ratio:** See: site coverage ratio.

**business park:** "High quality, low density environment, associated with excellent road and air communications and a metropolitan catchment area. Aimed at firms requiring a prestige image and high calibre work force. Mixture of manufacturing, product assembly and customization and sales and training functions" (Worthington, Chapter 3).

**business support center:** A new type of development, "the business support center is often located in the city center, often in a single office-type building, providing a mix of space and support services, research and even light high technology production" (Minshall, Chapter 4).

**cable management:** The effective handling of the increasing amount of cables and wires in the automated office, both at design stage and for day-to-day facility management purposes.

**card key lock:** Security locks for magnetically encoded cards that replace traditional keys.

**catalyst:** A term used in organizational theory to describe a type of person characterized as gathering information similarly to the "visionary" (see below), but making decisions based on the implications for all people involved, including him or herself, and aimed at making everyone happy. Catalysts will most likely be found in organizational cultures of the "developmental" type (Williams, Chapter 8).

**cathode ray tube (CRT):** A vacuum tube that directs electrons onto a fluorescent display screen to produce visible information.

**centralized building systems management:** Similar to: building automation systems (BAS).

**chip:** A thin wafer of semiconductor material (e.g., silicone, gallium-arsenide) used as a base for integrated circuits, and recently allowing ever more spectacular miniaturization of these electronic basic components. The building block of modern computers.

**clan culture:** Organizational culture where "it is believed that compliance flows from trust, tradition, and long-term commitment to membership in the system.... The individual's primary need is for attachment, affiliation, or membership" (Kimberly and Quinn, quoted by Williams, Chapter 8).

**coax (or: coaxial):** A type of cable combining several parallel (or coaxial) conduits within a single protective outer jacket, for such use as transmitting data and video signals.

**cognitive ergonomics:** A recently emerged branch of ergonomics that deals with the ways in which "the mediation aspects of human/technology interaction are designed to accommodate how people think and behave" (Springer, Chapter 9).

**configuration:** Refers to a group of pieces of equipment and machines that are connected and used together for a given purpose.

**convivial:** An adjective first coined by Ivan Illich to denote the appropriateness of the interaction between new technologies and people.

**computer:** An automatic electronic information-processing device.

**cooperator:** A term used in organizational theory to describe a type of person characterized as liking to deal with facts and direct experiences in light of the personal values of those affected by the information. Cooperators will most likely be found in organizational cultures of the "clan type" (Williams, Chapter 8).

**corporate culture:** "Is simply a way of thinking about how an organization goes about doing its business. A more formal definition is that it is the formal and informal patterns of behavior—the beliefs, values, philosophy, history and myths—which explain how things are done around here" (Williams, Chapter 8).

**corporate image:** The set of communication media (graphics, etc.) that is coordinated to project to the general public an image and distinguishable identity for a corporation's marketing purposes.

**covenants:** A set of rules, enforceable by law, that are designed to permit a local authority to control key aspects of a development, such as variances permitted, land uses, permitted activities, performance standards, architectural and engineering factors, and review procedures, etc. Covenants are a key to the success of high-technology parks.

**crowding:** A high density of individuals (animals or humans) in a restricted space that may cause stress by forcibly intruding into an individual's territorial distance.

**data:** Facts of information, especially as a basis for inference. Data can be processed by computers.

**data base (or: database):** Sets of interrelated data stored for ease of retrieval for particular purposes. In computing, the sets of data accessible to and operated upon by programs.

**data highway:** The central spine, or communications core, of an intelligent building (I.B.).

**decentralization:** The distribution of power. In contradistinction to: delocalization.

**decibel (Db):** A unit used to measure sound intensity. When A-weighted (DbA), it is weighted to suit the sensitivity of the human ear to different sound frequencies.

**delocalization:** The distribution of places and, hence, people. In contradistinction to: decentralization.

**deoxyribonucleic acid (DNA):** A substance whose molecules are in the shape of a double helix and that forms the basic material of genes.

**depreciation:** Term used in accountancy to denote the depreciating value over time due to natural wear and tear of a capital asset or piece of equipment, furniture, or machinery.

**design:** "Mental plan; scheme of attack upon; purpose; end in view; adaptation of means to ends; preliminary sketch for picture, etc.; delineation, pattern; artistic or literary groundwork; general idea, construction; faculty of evolving these, invention. Etymological origin: French: "desseing" (Oxford Dictionary).

**desktop:** See: personal computer.

**developmental culture:** Organizational culture where "it is believed that people will comply with organizational needs because of the importance or ideological appeal of the task. . . . In this dynamic view, it is believed that the individual's primary need is for growth, stimulation, and variety" (Kimberly and Quinn, quoted by Williams, Chapter 8).

**diagnostics center:** As part of building automation systems (BAS), the diagnostics center monitors critical environmental and performance features of a building and provides on-going information about the performance in-use of a facility.

**digital:** Information in code form (digits) that may be transmitted in the form of electrical pulses (on/off).

**direct digital control (DDC):** Control systems that operate by relaying information picked up by sensors to a computer-based automation system.

**dot matrix:** A matrix of dots generated by a printer, CRT, or VDT to form alphanumeric characters and/or images.

**downtime:** The period of time necessary to bring about changes in facility layout and building systems, usually with a cost tag in lost time and salaries. Minimizing downtime through flexible design contributes to improved productivity.

**Dynabook:** A "futures concept consisting of a single lap-size computer capable of holding all the information there is" (Stitt, Chapter 14).

**electronic blackboard:** An electronically sensitive chalkboard that scans the board contents and converts it into digital information that can then be printed out for hard copy use by the participants of a meeting, or sent down telephone line for visualization at a distance. Can also be useful as a complement to videoconferencing.

**electronic mail:** A form of person-to-person correspondence that uses electronic networks and media for transmissions.

**environmental psychology:** A relatively recent branch of psychology that focuses its inquiry on the relationship between people and their physical environment.

**epoxy coating:** Prevents moisture contact with reinforcing steel, hence spalling of cast concrete members in buildings.

**ergonomics:** Study of efficiency of persons in their working environments. From the Greek: *ergo,* work, and *nomos,* rules or laws. Other terms are used as almost synonymous, such as: human factors, human factors psychology, human engineering, or human factors engineering. "Ergonomics treats the human element as the primary determinant of workplace design" (Springer, Chapter 9).

**expert system:** In artificial intelligence, a computer system designed to incorporate expert knowledge of a given field, together with the procedures used by human experts in the field to access and apply such knowledge.

**facility (ies):** A generic term including "all aspects of the work environment" (Springer, Chapter 9).

**facility management:** The "amalgamation of attention—to the process, to the physical place, and to the people" (Williams, Chapter 8). More generally, the efficient management of facilities in use, and planning of new facilities, as an important aspect of the management and business plan of an organization.

**facsimile transmission (FAX):** The digital transmission of visual information via telephone lines.

**flat cable:** Cable made of copper conductor laminated in an insulating material to achieve minimal thickness. Laid under carpet tiles, it is used to connect transmission boxes to individual workstation outlets.

**flat screen display:** Terminal screens that are very thin, compared to traditional CRTs, with resulting minimal bulk, weight, and power consumption. The display material may be gas plasma, liquid crystal (LED), or thin-film electroluminescence. If and when they come to outperform CRTs for equal cost, they may become widely used, with numerous consequences for workstation design.

**fiber optics:** A laser-based relatively recent communications technology that transmits information in light form through "glass optical-wave guides," or "optical fibers," and enables very large amounts of information to be transmitted simultaneously. Fiber optics have low levels of loss or interference in signals, with high quality retained over long distances.

**flexibility:** The degree of adaptability of a facility to changes in user and/or technological requirements. Varying degrees of flexibility can be designed for, according to anticipated needs, with varying amounts of upfront capital investment.

**flexitime:** The flexible time tabling of working hours.

**footcandle:** A unit in the imperial system, used to measure illuminance and approximately equal to the illuminance projected by a candle at a distance of one foot. It is equal to 10.8 lux approximately, in the metric system.

**free cooling:** In energy-conscious building design, free cooling systems make use of unwanted heat for cooling purposes through condenser heat exchangers.

**functional information flow chart:** A chart akin to computer program flowcharts that represents the flow of information through a particular human/computer system.

**functional block diagram:** Diagrams representing the steps of a functional requirements definition.

**function allocation:** Phase in the human/computer systems design process when task functions are allocated either to the computer or to its user.

**functional requirements definition:** Accurate description of what must be done to accomplish the stated goal of the system.

**gallium-arsenide:** A material that has been increasingly used recently for the manufacture of integrated circuits instead of or in conjunction with silicone. It has remarkable conductivity characteristics.

**garage company:** A type of new enterprise similar to a "hard" company (see below) that started out developing an innovative product, with financing on a largely personal basis.

**glare:** Brightness or excessive contrast in illuminance that causes reduced visual acuity, discomfort, and eye strain.

**group computing (or: interpersonal computing):** A recently developed approach to computer systems designed for use by (problem-solving) groups, as opposed to individuals.

**hard company:** A type of new enterprise that tends to offer a standard range of products to an "anonymous" market.

**hard copy:** Computer output on paper or other physical support.

**hardware:** The physical components of a computer, including electronic, electric, magnetic, mechanical parts. In contrast to "software."

**Hawthorne effect:** An expression used to describe improvements in worker productivity that may result from a perceived increase in management's interest in its employees, even though actual improvements to work conditions, such as environmental ones, may not have taken place simultaneously. After a now classic series of experiments with lighting level in workshops carried out at Western Electric's Hawthorne factory by Mayo and Roethlisberger between 1927 and 1932.

**heat recovery:** Such systems are used in energy-conscious building designs to recover excess heat where it is generated (e.g., computer rooms) for reuse where it is needed (e.g., perimeter heat loss).

**hierarchical culture:** Organizational culture where "it is believed that people will behave appropriately or comply with organizational needs when roles are formally rated and reinforced by rules and regulations." (Kimberly and Quinn, quoted by Williams, Chapter 8).

**high-context culture:** A term coined by anthropologist Edward T. Hall to describe cultures, such as Japanese, where decision makers are routinely so aware of contextual information that they need little specific briefing information as input to any particular decision. In contrast to "low-context culture."

**High-strength concrete:** Concrete with ratings of 6000 pounds per square inch (psi) compressive strength and above.

**high technology:** A generic expression that broadly encompasses technological developments related to information and telecommunication technologies.

**high-technology industries:** "High-tech industries are usually defined as industries which are producing highly technical equipment that will be sold to consumers. High-tech industries might also be seen as those utilizing high-tech equipments to manufacture relatively conventional products" (Williams, Chapter 8).

**human engineering:** See: "ergonomics."

**human factors:** See: "ergonomics."

**human factors engineering:** See: "ergonomics."

**human factors psychology:** See: "ergonomics."

**incubator:** "An organization which helps with the development of new enterprises by providing them with space, services and counselling" (Y. Gasse, quoted by Davidson, Chapter 6).

**industrial motel:** Offers flexible rental space, but without services, contrary to an incubator.

**industrial park:** "Both standard and AAA, are clearly oriented toward traditional production, service and distribution and are not well suited to a wide range of high technology activities" (Minshall, Chapter 4).

**innovation center:** Multitenanted light industrial buildings, usually adjacent to a university, providing small units for firms growing out of university research groups. In some cases, an innovation center may have connections with an incubator.

**instruction:** The complete specification of an operation to be executed by a computer.

**integrated circuit:** A functionally complete electronic circuit contained on a base made of semiconductor material, such as silicone or gallium-arsenide. The building block of modern computers.

**intelligent building (IB)** (also: "smart building"): A generic expression coined to encompass all buildings incorporating a substantial amount of sophisticated information-rich systems that are digitally connected and centrally computer controlled. Two essential divisions are: centralized building systems management and shared tenant services (STS).

**interactive:** A computing application where, in contrast to one-way commands, the machine and the user respond to each other and exchange information, such as in question/answer or menu choices.

**interpersonal computing:** See: group computing.

**job aid:** Something to guide and help an individual's performance on the job, often complementing or replacing complete training, e.g., manuals, checklists, guidelines, handbooks, info graphics, etc.

**job description:** A general statement of what a job entails, including purpose, scope, duties, and responsibilities. Often used in labor relations.

**job enrichment:** Ways and means of increasing the scope of an individual's job through vertical integration, variety, and added responsibility. Often promoted to counteract dehumanizing aspects of Taylorism.

**job sharing:** The sharing of the responsibilities, time commitment, and rewards of a full-time job among several individuals.

**knowledge-worker:** Expression used to describe the increasingly numerous category of workers in "post-industrial" societies who rely on brain power, as opposed to the brawn power of production workers in industrial societies. Knowledge workers make considerable use of information technology and are generally well-educated professionals who expect high standards of quality of life and workplace environments.

**laptop computer:** Highly compact and portable computer that can run on mains or batteries. Uses flat screen display technologies (gas plasma, LED), and sometimes incorporates printers as well as modems for connection to networks. Their increased use in the workplace may considerably change the physical environment.

**laser sighting:** A very accurate and convenient leveling technique used on building sites that relies on laser beam technology.

**late-modern:** In architecture, a term coined by Charles Jencks to denote the school of architectural philosophy that continues today to develop and apply the basic tenets of the functionalist ideology of the modern movement of the early part of the twentieth century.

**light-emitting diode (LED):** A type of screen display lighting source.

**link analysis:** An ergonomics task analysis technique used to optimize arrangements of elements.

**local area network (LAN):** A communications network linking computer terminals and other electronic devices within an organization. There are three basic configurations for LANs: bus, ring, and star. The corresponding cable configurations are a major constraint on flexibility in space planning.

**low-context culture:** A term coined by anthropologist Edward T. Hall to describe cultures, such as North American, where decision makers are routinely so unaware of contextual information that they need considerable specific briefing information as input to any particular decision. In contrast to "high-context culture."

**lux:** See: "footcandle."

**master plan:** In urban planning and architectural design, a master plan sets out a broad framework for a proposed new development.

**mean:** Term used in statistics to describe the average value for a given sample of data.

**median:** Term used in statistics to describe the central point value for a given sample of data.

**meetings technology:** Expression used to describe information systems and media aimed at groups of users in meetings, rather than individuals.

**mode:** Term used in statistics to describe the most frequent value for a given sample of data.

**motion analysis:** An ergonomics task analysis technique that maps the discrete motions and corresponding times to perform a task or series of tasks.

**modem** (modulator/demodulator): A device allowing computers to communicate between each other via telephone lines.

**Myers-Briggs Type Indicator:** A questionnaire that consists of 126 items measuring an individual's preferences for perceiving the outside world and for making decisions about the information collected. It is most commonly used to measure the typology of Carl G. Jung.

**normal distribution:** Expression used in statistics to describe a sample of data where the mean, mode and median are all represented by the value for the 50th percentile. It is represented graphically by the so-called bell curve.

**obsolescence:** The deterioration over time of the suitability of a facility for its purpose, due to changes in user and organizational needs, or technological changes.

**occupancy sensors:** Sensors that detect whether a room or area is occupied. The information can be used for automatic lighting control, security, fire safety, HVAC control, etc.

**office automation (OA):** A generic expression used to describe the design and implementation of advanced computer systems for office purposes in organizations. A broad objective is to enhance productivity while improving work conditions and satisfaction.

**office of the future:** See: "automated office."

**office park:** "The office park tends to provide, not only for research and development, but also a wide range of office, light manufacturing and business support services as a result of less rigorous covenants" (compared to technology parks) (Minshall, Chapter 4).

**open office:** In space planning, the open office offers open floor area that can be laid out with furniture and equipment to suit organizational needs. As opposed to cellular offices, where partitions are used as space dividers.

**Organizations, Buildings and Information Technology (ORBIT):** The name of two major multisponsored studies, the first carried out in the U.K. (ORBIT-1), and the second in North America (ORBIT-2).

**overall performance** (also: total performance): For a particular facility, overall performance considers how the whole facility, rather than any particular part, component, or system, meets a particular need of its users.

**passive solar design:** In energy-conscious design, an approach to the conservation of energy through the passive, natural behavior of the building fabric. In contrast to "active solar design," which focuses on receiving, converting, storing, and using solar energy in electrical form.

**payback period:** Used in financial analysis of alternative investments to measure the length of time necessary to offset the initial outlay with anticipated returns on investment.

**percentile:** In statistics, a term used to designate the percentage of subjects in a given population whose measures are equal or less than a given subject's measures, e.g., for 5th percentile subjects, the measure of 5% of the population is less or equal to theirs.

**performance concept:** In building programming, the performance concept is used to specify the expected functional performance of a facility, as well as the corresponding technical performance criteria to consider and to measure.

**personal computer (PC)** (also: microcomputer): A small and autonomous computer with its own processing unit. Usually, it is dedicated to one individual and can be connected in network configurations.

**personal space:** In environmental psychology, an expression used to describe an individual's own private space (visual, acoustic, etc.), usually delineated by environmental markers. The type and size of personal space is individual and culture dependent.

**planning module:** A theoretical grid used in building design to align construction components such as partitions, floors, or ceilings.

**plasma** (or: Gas plasma display): A video display screen that uses gas plasma and gives a very clear image.

**plenum:** The space above a suspended ceiling when used for air circulation, usually return air, in an air-conditioning system. It is also usually used as a passage for ducts, cables, pipes, etc.

**poke-through (or: poke-thru'):** A method of cable management whereby cables are brought to a given floor through the ceiling of the floor immediately below, and brought up by poking through the structure with precut holes at predetermined (usually on a grid) points.

**Position Analysis Questionnaire (PAQ):** An ergonomic task analysis technique developed by McCormik, Jeanneret, and Mecham that consists of 189 work-oriented job elements constituting six major divisions of worker activity.

**post-industrial society:** Expression first coined by Daniel Bell to describe our present information-based society, as a successor to the industrial society where the manufacturing of products prevailed.

**post-occupancy evaluation (POE):** The measurement of the actual performance in use of a facility. POE permits the verification of predicted technical performance, as well as the continued evaluation of the performance of a facility in relation to changing organizational requirements. Criteria used for POE reflect facility programming criteria.

**power density:** A measurement scale that reflects the amount of energy (usually measured in watts) used per unit of floor area (usually square feet, in North America). Can be applied to overall energy performance, or to the energy performance of a particular system, such as artificial lighting.

**power management:** Systems that monitor the power consumption and cost of building systems and equipment for increased awareness of the causes contributing to utility costs and/or for billing purposes.

**printer:** Output device to produce a hard copy of a document. The two major categories of printers are impact and nonimpact printers. In environmental terms, the former may generate considerable noise.

**privacy:** Freedom from undesirable intrusions (e.g., looking in; looking out; acoustic, etc.). Environmental features allow the contrivance of adequate privacy for particular tasks.

**private branch exchange (PBX):** A private exchange connected to public telephone networks. Sometimes automated (PABX) to monitor and route communications (voice, data) to and from such networks, as well as within the organization.

**proactive:** Which plans action ahead of events. In contrast to "reactive."

**procurement:** The process that enables facilities to be procured for organizations.

**productivity:** Effectiveness of productive effort.

**program:** 1) In computing: a set of instructions for a computer to perform one or several functions. 2) In architectural design: a document that incorporates information on such topics as user needs, performance standards, budget, etc., as input to the design process. Hence "programming," "programmatic."

**programming** (or: "design programming" or "performance programming"): The identification of human needs and its translation into functional and technical performance requirements for an intended design.

**quality circle:** An approach to team management, initiated in Japan, whereby those workers closest to and most knowledgeable about a production process may contribute, and be rewarded to contribute, creative bottom-up suggestions for possible improvements and productivity gains.

**raised floor (or: access floor):** A floor construction method where the finished floor is raised well above (6 to 18 in) the structural floor level. The void is widely used in high-technology buildings to incorporate mechanical, electrical, and communications services.

**random-access memory (RAM):** In computing, the "live" memory that can be accessed directly during computation. In contrast to ROM, or read only memory.

**rational culture:** Organizational culture where "it is believed that people will comply with organizational needs if individual objectives are clarified and rewards are predicated on accomplishment" (Kimberly and Quinn, quoted by Williams, Chapter 8).

**reactive:** That which takes action in response to events. In contrast to "proactive."

**recreated reality:** The 3-D digital production of "impossible" images that picture mathematical functions that do not or cannot exist in the three dimensions of natural reality.

**riser:** Main vertical cable serving the floors of a building.

**science/research park:** "Development aimed at growing or established firms in research and development. Associated with a university campus, and jointly sharing amenities, research facilities, and know-how. Sites are 5 a or more." (Worthington, Chapter 3).

**sensor:** In an "intelligent building," a sensor detects data used by a centralized computer control system (e.g., heat, movement, light, etc.).

**shared tenant services (STS):** In "intelligent buildings," such services include a host of mostly electronic- and telecommunication-based functions that many individual tenants could not afford individually.

**silicone:** The most widely used semiconductor material for the manufacture of integrated circuits, or chips.

**site coverage ratio (or: "building to land ratio"):** The proportion of a building's ground floor area divided by the total site area.

**soft company:** A type of new enterprise based on individual contracts and development work.

**software:** The programmed instructions stored in a computer that made the "hardware" run. At their lowest level, such instructions are translated into machine code through binary digits (bits).

**specifications:** The written document or set of documents that complement working drawings to supply all the qualitative, quantitative, and contractual information needed to build a project.

**spin-off:** An often initially unforeseen activity, commercial venture, consultancy service, or product that results more or less directly from the main thrust of a scientific research program, such as located in a university, government research agency, nonprofit laboratory, or parent company. Spin-off companies are formed by individuals who draw heavily on scientific and technical knowledge used in their previous employment.

**spin-out company:** A type of new enterprise founded by someone who had previously worked in a large company on similar products or concepts, for which most of the R & D have already been done.

**stabilizer:** A term used in organizational theory to describe a type of person characterized as tending to value both scientific and direct experiences and searching for "truth" in developing procedures for action. Stabilizers will most likely be found in organizational cultures of the "hierarchical" type.

**stakeholders:** In organizational theory, refers to groups within organizations who share common stakes and objectives that may differ from those of other groups.

**standard:** "A rule for an orderly approach to a specific activity formulated and applied for the benefit and with the cooperation of all concerned" (ASTM definition).

**status:** In organizations, the status of individuals is frequently denoted by environmental features. Corporate cultures influence the existence and types of such visible demarcation.

**strain gage:** A type of sensor applied to key structural components of buildings to monitor their performance over time.

**surface reflectance:** Measures the percentage of incidental light that is reflected by a surface.

**supercenters:** "Buffer institutions" between organizations sharing common objectives and interests that serve the purpose of meeting interdisciplinary needs, allow for the sharing of resources, and provide flexibility in program development.

**superconductivity:** The property of certain materials (e.g., some ceramics) to offer little resistance to electrical current, a condition only achievable at near absolute zero temperatures until recently. The race is now on to devise superconductive materials at ambient temperatures.

**task:** In ergonomics, term used to describe what people do and how they accomplish their job duties.

**task analysis:** Phase in the human/computer systems design when human tasks are analyzed at an appropriate level of specificity to establish requirements for workplace elements and identify ways of optimizing performance from the human standpoint.

**Taylorism:** After Frederick W. Taylor, who was most responsible for developing in the early 1900s the principles of "scientific management" underlying the industrial engineering approach to the design of work. Basically, such principles consist in the "rational" fragmentation of work into task units that lend themselves to the most "efficient" use of people and machinery. Despite its monetary incentives, Taylorism is often mentioned as a major cause of alienation in industrial life. Job enrichment ideas attempt to counter such drawbacks by reintegrating tasks into coherent jobs that are satisfying to employees.

**technology park:** "Commercial development in a location with good academic institutions and an attractive life style to appeal to scientific and professional staff, low density, well-landscaped development of 100 a or more" (Worthington, Chapter 3).

**technology transfer:** The transfer of new knowledge from its producer(s) to its user(s) for didactic and/or commercial purposes, often occurring across disciplines or organizations. It is a major reason for strong linkages between universities and industry.

**telecommunication:** The transmission of meaningful information at a distance (e.g., by radio, telephone, television, etc.).

**telecommuting:** Term coined by Jack Nilles, of the University of Southern California, to describe the contemporary trade-off between commuting to work physically, and using telecommunications to work at home or away from the office.

**teleconferencing:** Conferencing at a distance by groups in two or more locations via telecommunications and electronic systems.

**terminal:** A peripheral piece of computing equipment linked to a central computer and through which input (keyboard) and output (display screen) of information occurs.

**territorial distance:** In most mammals, the radius of the perceived territory of an individual. Intrusion within territorial distance boundaries is generally perceived as threatening and stressful.

**thermal storage:** In energy-conscious building design, thermal storage allows energy to be temporarily stored, usually in the mass of the building structure and/or a water tank, for later use. Allows the evening out of thermal fluctuations (day/night, summer/winter).

**twisted pair:** Refers to the pair of copper wires that are twisted together for telephone and low-voltage communication lines.

**uninterrupted power supply (UPS):** Electrical power circuit, usually independent, that ensures that critical workstation computers receive a supply of power continuously.

**upgraded industrial estate:** "Many of the so-called European 'high-technology' developments are an upgraded marketing image applied to a standard industrial estate, where the same industrial warehouse sheds have a 'high-tech' facade applied" (Worthington, Chapter 3).

**variable air volume (VAV):** HVAC system controlled by monitoring the volume of fixed temperature air delivered with a local thermostat.

**vapor barrier:** A construction element that prevents internally generated moisture in vapor form from reaching outer-skin insulating material where it would otherwise run the risk of condensing and hence of reducing the thermal performance of such material.

**video display terminal (VDT)** (or: "visual display tube" or "video data terminal"): "A piece of computer equipment to communicate with a central computer for input (keyboard) and output (display screen). The screen may be a CRT, LED, or gas plasma.

**visionary:** A term used in organizational theory to describe a type of person characterized as the hard-driving achiever who values competence and logic in order to collect theoretical possibilities while subordinating human elements. Visionaries will most likely be found in organizational cultures of the "rational" type.

**weighting:** The assignment of coefficients of relative importance to the elements of a set of criteria used for scoring. Resulting scores are called weighted scores.

**wiring closet:** Closet designed to house major cables and related accessible equipment such as distribution boards, junction boxes, or patch panels.

**work:** Physical or mental activity undertaken to achieve a purpose. Often contrasted to play activity.

**working drawings:** In architecture, the set of drawings, mostly large scale, that gives full construction information about a project and is needed for work on site. Usually used in conjunction with specifications.

**workplace:** The physical setting where work activities occur.

**workstation:** An individual's work setting, including furniture, equipment, space, and building and other relevant systems. Often used to denote automated office settings.

**Xanadu:** A "futures" concept that consists of the categorization, and hence *pre*-creation, of all possible categories of information that can possibly exist.

**zone:** In HVAC systems, the word denotes a distinct space (a building or part of a building) for which identical environmental conditions (e.g., ventilation) prevail, controlled by one piece of equipment. The balance of zones needs to be adjusted where there is a change of internal layout and/or use.

# INDEX